Handbook of
Multichannel
Recording

Handbook of Multichannel Recording
By F. Alton Everest

TAB BOOKS
Blue Ridge Summit, Pa. 17214

FIRST EDITION

FIRST PRINTING—SEPTEMBER 1975

SECOND PRINTING—MAY 1977

Hardbound Edition: International Standard Book No. 0-8306-5781-9

Paperbound Edition: International Standard Book No. 0-8306-4781-3

Library of Congress Card Number: 75-20842

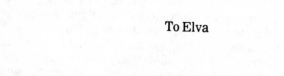

To Elva

# Preface

The recording industry is still slightly groggy from the effects of being hit by two major revolutions in the last few years—the coming of solid-state electronics, and the swing to multitrack techniques. It is difficult to envision the modern multichannel audio control console and the multitrack recorder without the space-saving and other advantages of transistors and integrated circuits. A third revolution has already started—digital audio processing and the glittering promise of computer science bringing automation to help in the ever growing complexity of multitrack operation.

But what is multitrack recording? It's all a matter of where and how the sound mixing is done. With a single microphone picking up music from a band, the sounds are irrevocably mixed as they fall on the microphone. Another approach is to dismember the band and pick up the sounds of each fragment individually and store them on individual tracks of a multitrack magnetic tape. The mixing is done later. Long after the musicians have left and calm has once more settled on the studio, the tedious job of creating something fresh and vibrant out of all those separate tracks begins. It can go on for days or weeks. Some tracks may be erased and a lone musician called into the studio to replay that part, listening to a temporary mix of the other tracks with headphones as he plays. Artificial reverberation may be added, balance between the bass and treble regions adjusted, relative levels trimmed, the tonal character of some instruments completely changed by exotic electronic gizmos. Little by little the pieces are put together with a completely new spatial relationship that may have nothing to do with the physical arrangement of the musicians in the studio, as stereophonic and quadraphonic spaciousness is added. The resulting composition bears little resemblance to the sounds a single microphone would have picked up in the studio. It's not only different, it's orders of magnitude better; and the exploitation of this new technique has only begun.

Multitrack recording, the basis of the recording industry today, is founded on a fusion of the esthetic and the technical. The burden of this book is the technical underpinning of modern recording, to advance the knowledge of the technical tools of esthetic expression. An understanding of these tools is important to those engaged in it from the business, musical, or technical perspective.

This volume is a handbook of the broad aspects of the practice and principles of multitrack recording for those working in the industry as managers, producers, directors, or recording engineers. As the average hi-fi buff has sufficient technical background to grasp the concepts presented in this book, it should be of special value to those interested in getting into the recording game. To encourage the development of new talent on both sides of the microphone, the last chapter is devoted to the budget multitrack operation.

F. Alton Everest

Contents

The Rise of Multitrack Recording

The wire recorder of the 1930s generated a considerable amount of interest, not so much for what it could do of itself, but for the promise it held for the future. In it an entirely new principle of recording was presented to the world, giving hope of release from the fragility of the wax cylinder and the scratchiness of early discs. This promising infant burst upon the world in swaddling clothes of tangled wire and some very impressive distortion. But the principle of recording a signal by means of invisible variations of magnetic flux in a magnetic material—instead of gouging out material from the surface of a semisoft cylinder or disc—opened up a whole new vista for the future. The keener scientists and engineers saw in this novel idea the promise of the long sought medium for recording high fidelity sound, sharp pictures, and other forms of information. During the same period of development the technique of cutting master discs and pressing high quality records grew into a major industry and the disc medium was widely adopted by the public.

With the ingenious addition of high frequency bias, magnetic recording quality took a quantum jump. Magnetic recording was now on the royal high road to success, accelerated by the introduction of an iron oxide coating on a plastic tape, equalization of response, improvements in heads, improved constancy of tape motion, etc. Today the thin magnetic coating medium appears in fields as diverse as home cassette recorders, video recording, and for the storage of data in computers.

In this book our interest will be focused on systems which end in the recording of sound on tape, from the ubiquitous ¼ inch width to 2 inch width, accommodating 8, 16, 24, or even 40 tracks of various components of the same acoustic event laid down side by side. We shall see how this introduces a new standard of flexibility accompanied by a host of new problems.

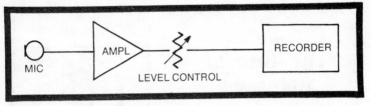

Fig. 1-1. Basic elements of a single-track magnetic recorder.

Types of Recording Systems

In Fig. 1-1 the sound signal picked up by a microphone is amplified and fed to a magnetic recording mechanism through a gain or *level* control of some sort. (We also need some device to help us to determine signal level, but we'll neglect all such peripheral items as we concentrate on the basic principles.) The system of Fig. 1-1 is a monophonic (single-channel) system characteristic of all early tape recorders, amateur and professional alike, and invariably they recorded over almost the full width of the magnetic tape. A variant of this system which soon came into the amateur market was the half-track machine, which utilized only half the tape so that recording time could be doubled by turning over the full takeup reel as it was placed on the supply reel spindle. Note that any system incorporating a reversal of tape motion by turning over of reels rules out editing of the tape and hence its use is limited to the consumer field.

The professional system of the same period, used for monophonic recording and in radio broadcasting in many parts of the world even today, is illustrated in Fig. 1-2. Many microphones, disc reproducers, or other inputs are handled by a multichannel console which combines the signals from the various inputs into a single composite signal which is usually recorded over the full width of a magnetic tape. Program equalization is applied to the individual channels, the composite signal, or both.

Although stereophonic techniques were demonstrated on a laboratory basis in the early 1930s by Bell Telephone Laboratory personnel and others, stereo did not hit the consumer market until the availability of tape recorders in the 1950s. Figure 1-3 illustrates the common hi-fi stereo recording system. This requires a two-track recorder. In consumer recorder—reproducers the two tracks for the forward direction (tracks 1 and 3) are interleaved with the two tracks for the opposite tape direction (tracks 3 and 4), and the so-called four-track machine results.

12

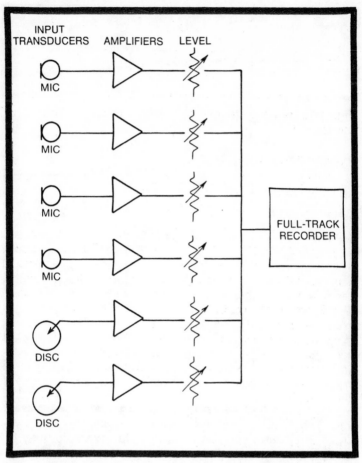

Fig. 1-2. Multichannel mixer and single-track recorder system still widely used in monophonic broadcasting.

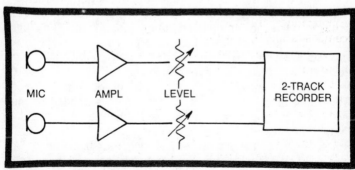

Fig. 1-3. Basic elements of stereophonic recording system.

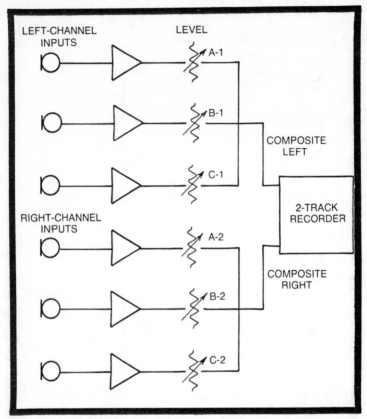

Fig. 1-4. Multichannel stereophonic mixer feeding two-track recorder. Faders A-1 and A-2, for example, may be ganged.

The professional stereo recording setup was basically very similar except that only a single tape direction was used so that editing of the tape was possible. The stereo recording console also accommodated many inputs in the highly simplified sketch of Fig. 1-4. In a stereo console, level controls A-1 and A-2 would be mechanically ganged to control the signal level from a stereo pair of microphones.

The system shown in Fig. 1-5, used widely for the recording of popular music, is now being extended to more serious music as well. Here, each channel may feed an individual track on the magnetic tape. Or a group of microphones on the drums, for instance, may be combined and recorded on a single track. Sixteen tracks are common in such recorders and many more tracks are technically feasible by crowding more recording heads into the limited tape width or through electrical interlocking of two or more recorders. There seems to be no electronic limit to the number of channels of an audio control console, although there is growing concern for the human factor of managing several hundred knobs and effectively riding gain on so many channels. However, the improvement in flexibility, creative latitude, and overall economy cannot be denied.

The storage of the original recorded material on, say, a 16-track, 2-inch magnetic tape is only an intermediate step. These tracks must be combined to obtain, for example, a two-track stereo magnetic master to be used in cutting the

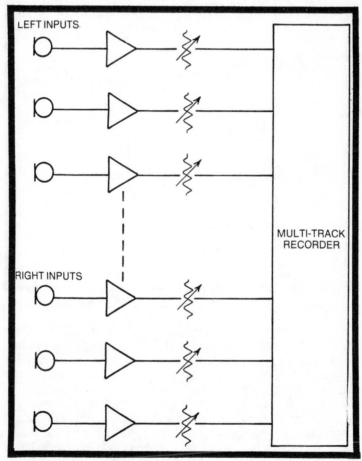

Fig. 1-5. Basic elements of modern multichannel mixer feeding a multitrack recorder.

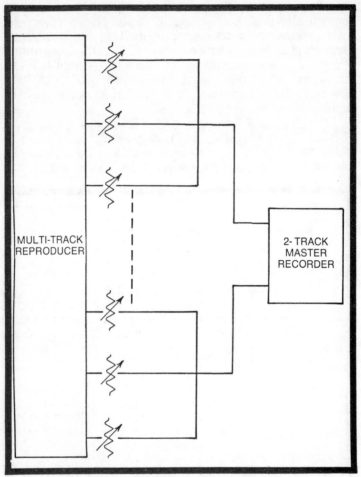

Fig. 1-6. Basic elements of modern stereo mixdown.

lacquer master disc in preparation for pressing production discs. This mixdown process is indicated in highly simplified form in Fig. 1-6. Only the channel faders are shown for adjustment of relative levels which could very well be the faders of the same console used in recording. But this console would also provide equalization in each channel for adjustment of frequency response, panoramic potentiometers (*panpots*) for placing the signal of each channel in the stereo field, reverberation (echo), and other effects, to name but a few. All of the flexibility of the multichannel/multitrack system comes into focus in this mixdown operation, and this is the time the producer molds the original tracks into a complete, salable composition. This is the time for esthetic manipulation of the original tracks; artistry every bit as great as that of the performers may be expressed at this mixdown stage.

Although Fig. 1-6 shows a stereo mixdown, the original multitrack recording may as readily be mixed down to a monophonic or quadraphonic format. With the segmentation of the sounds of the original performing group onto the multitrack tape, a variety of options are offered as to how the parts are reassembled. The uninformed skeptic might marvel at the spectacle of recording on 16 tracks and then releasing on two. Proponents point out that this shifts the major artistic decisions from the frenetic recording session to the relative tranquility of the mixdown room.

Music of the younger generation ever rocks and rolls on to new commercial heights. Many changes in recording equipment and techniques have appeared to serve this burgeoning industry. The multichannel console and the multitrack magnetic recorder demanded by popular music groups are at the very heart of this revolution. It cannot be denied that these young musicians have provided a great impetus toward the multichannel/multitrack practice so common today; they have, in fact, introduced many of the techniques themselves.

SEPARATION RECORDING

The many recording channels now available may be used (a) for multimicrophone recording of a group performance, or (b) by assigning a microphone to each individual performer or small group of performers who are part of the group effort. In (a) the final mix is accomplished during the performance. In (b) the separately recorded tracks are later brought together in a remix or mixdown session. Method (a) is typical of the way a symphony orchestra, for example, was recorded in the old days and some of the finest recordings have been made this way. The same orchestra can be recorded by acoustically isolating the strings, woodwinds, percussion, brass instruments, and soloists from each other in the recording session according to method (b) and then combining them in the mixdown.

The procedure followed in putting the multichannel gear to work determines whether the *buncher* or the *separatist* philosophy of recording is applied. The buncher prefers to mix components of the group effort live with natural interaction between performers. The separatist would rather record the performance of each individual (or small groups of

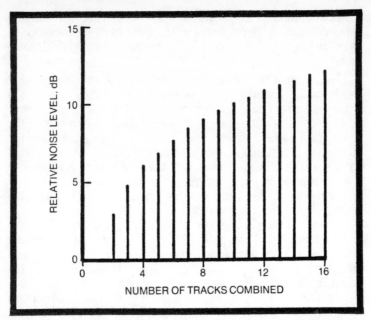

Fig. 1-7. Noise buildup in combining material recorded on separate magnetic tracks.

individuals) on separate tracks and do the combining at a later time, probably making many changes in the quality of each track in the process.

In the separatist philosophy the final recording is more "the creature of the control room and less the documentation of an acoustical event," as John Eargle put it[1]. This creates a problem: The recording technicians may exercise greater artistic responsibility during the mixdown than the musical director during the recording! Involvement of the musical director in the mixdown would tend to resolve this dilemma, but it is the rare traditional musical director who is emotionally capable of first presiding over the dismemberment of his group and then spending long hours putting things back together with radical changes. However, the voluminous output in the popular music field demonstrates that at least musical directors of popular works are capable of such schizophrenic performances and are often deeply immersed in the technical side of recording as well.

The Debits of Separation Recording[2,3]

Separation recording inescapably means multitrack, and multitrack means higher noise. When two tracks having equal noise levels are mixed together, the resulting noise level is increased 3 dB. Mixing 16 tracks down to mono results in a combined track noise 12 dB higher than the original tracks, as shown in Fig. 1-7. Note that these are not *absolute* values, but rather *relative* measurements. If the noise of one track is measured at −80 dB, then the noise voltage of 16 tracks would be 68 dB below the 0 dB reference point.

Separation recording tends to reduce the overall dynamic range of the system. All other factors being equal, reproduced signal level from a given track is a function of the width of the tape track; and in multitrack recorders the width of the gap of each head must be reduced to accommodate many tracks on a given width of tape. The combination of decreased reproduced signal level and increased noise means that the signal-to-noise ratio and the dynamic range of the system are reduced. This deterioration of dynamic range can be helped by the application of effective but expensive noise reduction systems. (Dynamic range is defined as the total usable range in audio volume, between noise at the one extreme and full undistorted volume at the other.)

Greater crosstalk is another result of crowding more and more tracks onto a tape of given width. Let us consider two instruments recorded on adjacent tracks. If the studio acoustic separation is only 10 dB, crosstalk in the recording and reproducing head will be no problem. If these two instruments are recorded at different times, the head crosstalk may become noticeable, but as long as both are components of the same musical selection, there should be no trouble because modest amounts of carryover between tracks is desirable, creating a feeling of greater expansiveness. If, however, the two adjacent tracks carry unrelated material, crosstalk may be a serious problem.

Confusion of artistic responsibility is often a byproduct of separation recording. There is no such thing as recording a selection with that old-time feeling of elation in director, performers, and recording engineer alike that it is a job well done. Rarely is anything finished until some of the tracks have been redone, others picked up later in overdub sessions, and many hours have been spent in mixdown. It is a rare, persevering, omnipresent musical director who supervises all this; and if he just checks results here and there, compromising of his goals is inevitable.

The creativity of musician, recording engineer, and director tend to be subjugated to mechanical things as creation of new and unusual sounds supplants the goal of faithful reproduction of the original sounds in the studio.

As musicians are separated from each other physically and acoustically, something tends to be lost in the music in the effect the musicians have on each other. The intangible "something" that makes a group successful is undermined to a certain extent. Playing to a cue track over headphones is something less than having one's fellow performers an armlength away. Physical separation, extremely dead studio acoustics, opaque baffles, and isolation booths achieve channel separation all right, even to the extent that the musicians often cannot hear one another.

Multitrack usually means two-inch tape, which brings with it certain problems of stretching and deformation. In videotape recorders, vacuum ports are used to hold the tape in intimate head contact, but not so in multitrack audio recorders. The deformation problems increase as the tape is subjected to repeated passes through the machine; separation recording is notorious for the number of runs required and the excessive wear to which the original tape is subjected.

The Merits of Separation Recording

The advantages of multitrack recording may be summed up in one word: flexibility. The spectrum of this flexibility is broad indeed.

Multitrack recording offers the ability to conquer space and time by building up a group effort piece by piece. For example, an acoustic guitar and drums could be recorded on separate tracks on Monday when these performers are available. Then, in overdub sessions, other guitars might be added on Tuesday, bass on Thursday. The tape could even be shipped across the country to pick up a big name vocalist a few days or weeks later. In each overdub session, headphones carrying a temporary mix of what has been recorded supply the performer with necessary timing cues.

A huge advantage of separation recording is control—the freedom to adjust the relative level of each track in the mixdown process to create the desired balance in the end product; the freedom to equalize each track according to its individual needs or to achieve special effects; the freedom to alter the quality of each track to achieve unusual effects. Many of the electronic effects used in synthetic music generators can be applied at this stage.

Separation recording makes possible the addition of reverberation (often incorrectly called *echo*) to any track in any degree desired to compensate for dry (dead) studio acoustics or to achieve special effects.

The separation philosophy allows the introduction of special effects such as *fuzz* and *flanging* or other phasing effects to any individual track (discussed in detail in Chapter 7).

Perhaps the most important advantage in these times is economy. The technique of separation recording has proved a real money-saver for perfectionist music directors who at one time had to have whole crews and musical groups stand by for retakes when a single performer flubbed a passage. With separation, the big performance decisions can be deferred to the relatively low pressure mixdown stage rather than having to make them during the emotionally charged recording session with many expensive artists standing around.

Fortunately, there are ways to minimize all of the disadvantages of separation recording listed and reduce their effect, sometimes to negligibility. For example, the problem of greater noise (tape hiss and other background noises) can be met head-on by the standard practice of recording at maximum level. In Fig. 1-8A the old-style *premixed* recording is illustrated. Five instruments are balanced out and the relative levels, represented by lines of varied length, are just what the musical director wants; they are mixed this way and recorded in their final form on the spot. The shorter the line, the lower the instrumental level and the closer to the channel noise level for that instrument.

Even if the playback system were completely noise-free (impossible, but a suitable example for our purposes), it is easy to see that one of the instruments is but 30 dB above the noise at full volume. And if that instrument is playing a solo passage that calls for a very low-level bar, the noise will be disconcertingly apparent on playback. In the example shown, the low-level instrument's level is not even 40 times greater than the inherent tape noise.

In separation recording, illustrated in Fig. 1-8B, all instruments are recorded at the greatest level possible without distortion, with the result that all signals are — in this example—some 60 dB above the noise. In terms of level, this spread means the intended signals are 1000 times stronger than the noise signals of the individual tracks!

When the taped master is ultimately played back to cut the master disc, the levels of the individual instruments can be controlled by manipulating the level controls of the respective tape tracks; and when this is done, the noise levels drop proportionately. Of course, a whole new noise level will be encountered because of the disc's limitations, but at least the noise contributions of the tape will have been neatly sidestepped.

Separation recording makes possible another attack on the noise problem based on the principle that the narrower the

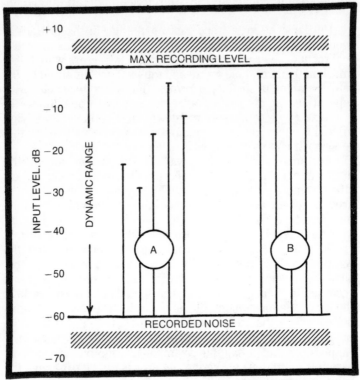

Fig. 1-8. When separate channels are mixed before recording, as in A, the overall signal-to-noise ratio on many of the channels is poor. In modern multitrack recording, all tracks are recorded at high level (B), resulting in an improved signal-to-noise ratio in the mixdown.

accepted frequency range, the lower the system noise. Each instrument has energy distributed over its own characteristic frequency range and each equalizer can be adjusted to fit the instrument feeding that channel. For instance, the violin sound has practically no energy below 100 Hz, so adjusting the equalizer to cut off everything below this frequency would help the noise just that much for that channel. The bass has significant energy down to about 30 Hz but less in the high frequency range, making it desirable to roll off at about 30 Hz on the low end and cut off at 3 to 5 kHz on the high end. Thus restricting the channel's passband reduces the noise in that channel which contributes to a general reduction of overall noise level. If violin and bass and a conglomeration of other instrument sounds are being picked up by a single microphone, such fitting of bandwidth to specific signal requirements is not possible.

This noise advantage is an inherent part of the separation recording technique, and tends to offset the problem of noise increasing with the number of tracks. In spite of all this, noise reduction systems are still needed and are almost universally used.

Another powerful argument supporting separation recording is that this is what the customer demands today. Even though he may have little logic and much emotion behind his demand, he pays the bill and thus calls the tune. Woe to the studio offering only four tracks when a potential customer appears demanding sixteen! Pitifully, the number of tracks, not the expertise of the operators, is often the standard by which a studio's capability is judged.

Growth of the Recording Industry

In the Billboard International Directory of Recording Studios the date each studio was established is listed.[4] By tabulating these figures, some very interesting information comes to light on the growth of the recording industry in the United States. In Fig. 1-9 the number of *new* studios established in each year is shown in bar chart form. The astounding growth during the decade 1964–1974 explains the present great interest in modern recording techniques and in recording studios. Figure 1-10 presents the growth of the

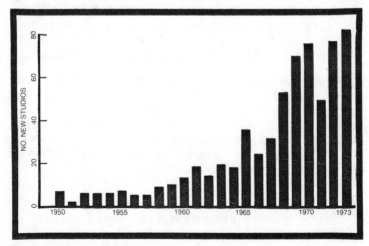

Fig. 1-9. The number of new studios established each year in the United States. Data taken from listings in Billboard International Directory of Recording Studios.

23

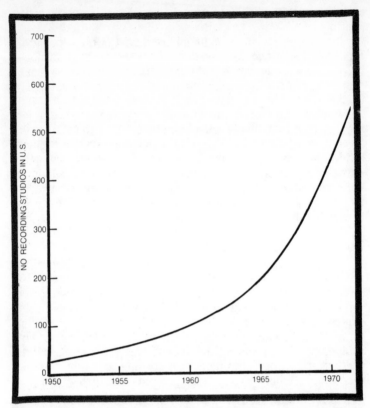

Fig. 1-10. Cumulative growth of number of recording studios in the United States as determined from listings in Billboard International Directory of Recording Studios.

number of recording studios in the U.S. in cumulative form. During the 1950—1960 decade the growth rate was about 8 new studios per year on the average. During the 5-year period 1968—1973 this growth rate shot up to about 70 new studios per year. This graph errs on the low side as it is based on the year of establishment of each studio. Many listings in the Billboard directory do not include this information.

Record Production

Before we dive into the details of recording, it is well for us to view the overall process of producing a usable, salable consumer product. The product may be a special production for radio or television, but this accounts for only a fraction of the recording work going on. The bulk of recording activity is directed toward producing phonograph records for the home

24

market. For the time being, at least, disc records far outshine other forms of release in popularity, although high speed duplication of cassettes, 8-track cartridges, and quarter-inch tape is growing fast; and the quality of these products and their reproducers is steadily improving.

A simplified diagram of the multitrack recording operation is shown in Fig. 1-11. Many inputs are fed into the console and many outputs go to the multitrack recorder. A console designated 24/16 accommodates 24 inputs and 16 output signals. In recording a musical group, microphones feed most of the input channels of the console. Sometimes the demands of the situation require more inputs than the board has, in which case a certain amount of premixing is done with outboard mixers. Premixing is also a part of the console manipulation to squeeze 24 input channels into 16 output tracks, for example. It is most important to note that anything premixed and recorded on a single track cannot be unmixed later and that a certain amount of mixdown flexibility is sacrificed with every premix. In practice, however, this sacrifice may not be too great as certain sounds are combined during recording. For instance, five microphones picking up audience reaction might very well be mixed on the spot, or a group of similar musical instruments could be logically premixed with minimum sacrifice in flexibility.

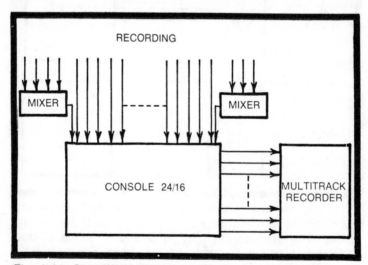

Fig. 1-11. Simplified diagram of typical multitrack recording operation. Premixing can increase the number of inputs accommodated above the number of channels of the console.

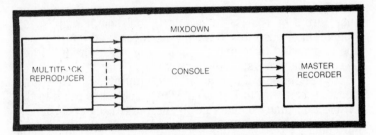

Fig. 1-12. In the mixdown (remix) process, the multitrack recorder becomes a multitrack reproducer; the same multichannel console can be used.

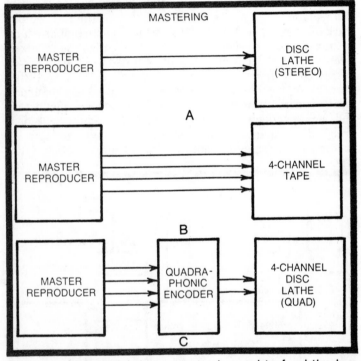

Fig. 1-13. The mixed down master is used to feed the lacquer cutting lathe for mono, stereo, or quadraphonic releases. The process depicted in C is descriptive of both matrixed and discrete quadraphonic discs: For matrix, the two disc channels are phase-varied to create a form of modulation that may be decoded to form a pair of extra channels; for discrete, the two disc channels carry audio and supersonic signals that may be decoded to form the four individual channels.

The mixdown, or remix process is diagramed in Fig. 1-12. The multitrack recorder is also a multitrack reproducer or playback machine and its tracks are fed into a multichannel console. The same recorder and console used in the recording stage may be used, depending on the size, organization, and workload of the studio. The recorded tracks fed into the console in the remix are combined into four tracks for quadraphonic release, two for stereophonic, or one for monophonic. These are called master tracks and the machine they are recorded on is called a master recorder, which is usually a quarter-inch machine of high precision.

Although not treated in detail in this book, the mastering process depicted in Fig. 1-13 is important for our consideration because the remixed product of the recording operation must be in such a form as to preserve the utmost in quality as it is passed on to the mastering laboratory. Although the mastering engineer is capable of adding touches which could improve the final product, he is often hindered by lack of data or slipshod techniques. The mastering engineer's main task is to optimize the tracks for frequency response, noise, and distortion *in the finished product* as the tracks feed the driver of the master disc cutting lathe.

For a quadraphonic master to be cut, either a four-track master or a matrix encoding step is necessary; for cutting a stereo record from a two-track master, no encoding is necessary.

REFERENCES

1. Eargle. John. *Equalizing the Monitoring Environment.* Jour. Audio Engr. Soc., Vol. 21, No. 2, March 1973 (p103).

2. Woram. John. *Anyone for Two-Track?* **db.** The Sound Engineering Magazine, Vol. 4, No. 2, February 1970 (pp 16—17).

3. Alexandrovich, George. *Multi-Channel Recording—Why?* **db.** The Sound Engineering Magazine, Vol. 3, No. 3, March 1969 (pp 4, 6, 8), and Vol. 3, No. 4, April 1969 (pp 4, 6).

4. Billboard 1974 International Directory of Recording Studios.

2 Management for Track Separation

The flexibility and success of multitrack recording stands or falls solely on the ability to provide tracks separated enough from each other to allow reasonably independent handling of each track in the subsequent mixdown.[1,2,3] An interesting case out of recording antiquity is the brilliant manner in which Walt Disney Studios recorded the Philadelphia Philharmonic Orchestra for the film *Fantasia* in 1938 and 1939. This is not a strict example of modern multitrack recording, but to achieve certain spatial effects in the sound for this film, the audio engineers approached the problem on a very modern looking multitrack basis. The orchestra was recorded on eight separate tracks as follows:

TRACK	CONTENT
1	violins
2	cellos
3	bass
4	violas
5	brass
6	woodwinds and tympani
7	balanced mixture of 1–6
8	overall orchestra with distant pickup.

The conductor, Leopold Stokowski, was placed at the center of a cluster of polygonal cells[4] as shown in Fig. 2-1. The cells were made of inch-thick plywood with rear walls covered with absorbent material. The output of the microphone of each cell was recorded photographically on its own 35 mm variable-area film recorder. These tracks were mixed down to

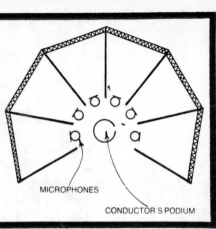

Fig. 2-1. Arrangement used in historic application of multitrack technique in recording the Philadelphia Philharmonic Orchestra for the film Fantasia in 1938-39. (© Walt Disney Productions)

MICROPHONES

CONDUCTOR'S PODIUM

three optical tracks and a fourth carried level control signals for each of the three sound tracks so that the audio could be made to follow the action on the screen. Loudspeakers placed behind the screen, to the left and right of the audience, and overhead, together with some judicious loudspeaker switching, made possible the well known effects in the original roadshow version of this film. Present releases utilize fewer loudspeakers with a resulting restriction in effect. The orchestra's main job was to enhance the mood and action of the several animated musical numbers in the film, rather than appearing on the screen. This gave the engineers full freedom to move the sounds around; in this regard, Fantasia sound followed modern techniques quite closely as the final effect was obtained by manipulating the separate tracks (Figs. 2-2, 2-3, 2-4).

ACHIEVING SEPARATION

Just how much separation is required in multitrack recording? The engineers of Japan Victor Company (JVC) decided on an intertrack separation goal of 15 dB as an average figure in the design of their studios. Others strive for 20 dB or more, depending on the circumstances and the nature of the group being recorded. In terms of microphone voltage or sound pressure, a separation of 15 dB represents a ratio of only 5.6 to 1 and 20 dB a ratio of only 10 to 1. Of course, it is the relative loudness to human ears that really counts; and this bears a complex relationship to readily measured voltages and pressures. Also, we must recognize that there is usually compatible sound on the next track and all we are really asking of this 15 to 20 dB separation is to allow freedom to establish relative dominance between performers. It is not like crosstalk in other communication channels, in which an

Fig. 2-2. Panpots requiring six hands, bicycle sprockets and chain used in the pioneering multitrack mixdown of sound for Walt Disney's late-thirties classic, Fantasia.(© Walt Disney Productions)

Fig. 2-3. The "Six-Headed Monster," a ganged arrangement of six Moviola sound heads and one picture head used in the editing of Fantasia. (© Walt Disney Productions)

Fig. 2-4. Walt Disney explaining his ideas for presenting Mickey in the role of the Sorcerer's Apprentice to Leopold Stokowski and Deems Taylor. (© Walt Disney Productions)

entirely unrelated conversation might interfere with the program material of an adjacent channel.

SEPARATION BY DISTANCE

Let us take a musical ensemble of some sort which can be broken down into five instrumental or vocal subgroups, as shown in Fig. 2-5. A single microphone at position A would pick

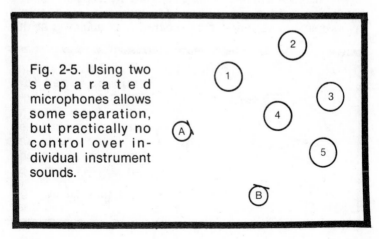

Fig. 2-5. Using two separated microphones allows some separation, but practically no control over individual instrument sounds.

up a mixture of sounds from all five sources. There would be a certain amount of level difference resulting from small differences in distance and differences in acoustic output of the various sources. There is nothing that can be done with whatever composite signal exists, however. They are mixed together for time and eternity and nothing can be done to the sound from one source without affecting the sound from all the other sources as well. Practically speaking, there is no separation since but one microphone (one audio channel) exists. No matter how many speakers are employed in the reproduction, each speaker will reproduce essentially the same material.

Suppose we now add another microphone at position B, and use it to feed another tape track. Now we have introduced separation. Instrument 1, being physically closer to microphone A than to microphone B, will be reproduced as a greater percentage of A's overall volume than of B's. Instrument 5 will make up a greater percentage of B's overall volume than of A's. Instrument 4, about the same distance from both mikes, will be reproduced by both at about the same level, which translates to a similar positioning on playback. This form of achieving separation is extremely basic, and very little control is offered in later tailoring, since any effort to vary the level of any given instrument will affect the level of all instruments recorded on that channel. Instrument 2, for example, is situated about the same distance from both mikes; on stereo playback this instrument will appear to be at position 4 (or just behind it if the amplitude is significantly lower than instrument 4).

The next step toward achieving separation between the five sources would be placing a separate microphone close to each source as in Fig. 2-6, and feeding each to a separate recording track. Only partial separation is achieved at this

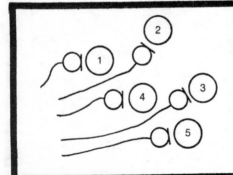

Fig. 2-6. Separate microphones for each source do not achieve much separation when the sources are close together, since each mike picks up signals from adjacent instruments.

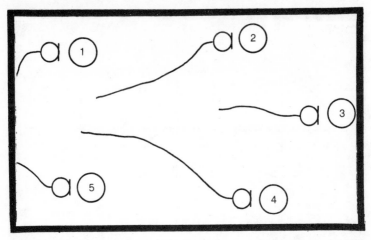

Fig. 2-7. Microphones close to their respective sources plus physical separation of sources gives some really usable track separation.

point because of the short distance between adjacent instruments and microphones. Separating the sources, each still with its own close-in microphone as in Fig. 2-7, begins to yield something that approaches satisfactory track separation. If this physical separation places some sources near reflecting surfaces, problems may be encountered; but these can be minimized by using a dead acoustical environment.

Obviously, the content of any two tracks would have some effect on the separation required. A vocal soloist may require greater separation from the drums than from the guitar to give maximum flexibility in handling them later. Beyond microphone and subgroup placement are other things affecting track separation: microphone characteristics, baffle placement, and studio acoustics, to name but a few.

In summary, then, we have seen that separation between instrumental or vocal sources is affected by (1) the closeness of each microphone to its source, (2) the distance between sources, and (3) the relative output level of the different sources. There are other ways we can improve separation besides separating sources and moving microphones closer to the sources:

- Utilization of microphone directivity
- Use of baffles and barriers between sources
- Juggling of acoustics of recording space, including reflectivity of surfaces

33

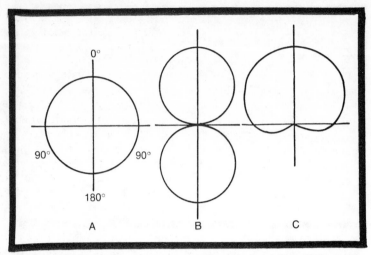

Fig. 2-8. The three common microphone directivity patterns: (A) omnidirectional, (B) bidirectional, and (C) cardioid.

- The use of electronic devices
- Use of contact transducers

And we should at least recognize the extreme forms of separation made possible by the overdubbing technique of recording different instruments or vocalists at different times or in different studios.

MICROPHONE DIRECTIVITY

An omnidirectional microphone, Fig. 2-8A, as its name implies, is equally sensitive to sounds arriving from almost any direction and offers no discrimination against unwanted sounds from adjacent sources.

The bidirectional microphone of Fig. 2-8B, having a sensitive lobe forward and another backward, offers some help in achieving separation. The bidirectional microphone is typically of the pressure gradient or ribbon type and responds to the difference in pressure at two successive points along the path of the sound wave. If the microphone is placed sideways to the path of the sound, the pressure is the same at these two points and there is zero response. The microphone is thus responsive to front and rear but dead to sound arriving from each side. These side nulls can be aimed at sources from which pickup needs to be reduced. However, care must be exercised to control what the rear lobe aims at. The voltage produced by sound arriving from the rear is 180° out of phase (opposite in polarity at any given instant) with that arriving from the front. The figure-8 bidirectional microphone is often used in the natural stereo pickup in which two of these microphones are mounted at right angles to each other.

By combination of an omnidirectional microphone element and a bidirectional element, the cardioid (heart shaped) pattern of Fig. 2-8C is obtained. Some cardioid microphones actually incorporate both ribbon and moving-coil elements; others achieve the same effect through acoustic networks. The outputs of the two components add for sound arriving from the front, cancel for sounds from the rear, and partially cancel at the sides. The cardioid microphone is essentially unidirectional, but the sensitive lobe is quite wide. For practical purposes its width is usually considered to be 100−120°.

These three basic microphone response patterns[5], or variations of them, can be used in the quest for adequate track separation. There is another type, the highly directional microphone, usually characterized by large size, which finds only limited use in multitrack work. The forward lobe is sharp at high frequencies, but generally degenerates into broader response at low frequencies.

BAFFLES

There are situations in which even the physical separation of the sound sources of Fig. 2-7 does not provide sufficient acoustic separation between tracks. For example, let us say that source 2 is a thin-voiced vocalist, source 3 an amplified guitar, and various reasons have dictated the expediency of the indicated arrangement. Placing a baffle between sources 2 and 3 as in Fig. 2-9A offers some shielding of microphone 2 from 3. In placing this baffle we should be alert to the possibility of a bounce of source 3's sound to microphone 4, thus creating another problem. If such a problem were to arise, baffle A could be placed differently or a baffle with absorbing material on the side of source 3 could be employed. Because of the weakness of the vocalist's output we wouldn't expect too much dilution of the microphone 3 signal. For the sake of illustration, let us make source 1 the drums (Fig. 2-9B). They are too loud at all the other microphones, so baffles are placed to minimize this problem. More baffles could be deployed as needed to improve track separation.

This example has considered only the direct component of sound; but virtually every studio, no matter how dead, has walls, floors, and ceilings that may reflect enough sound

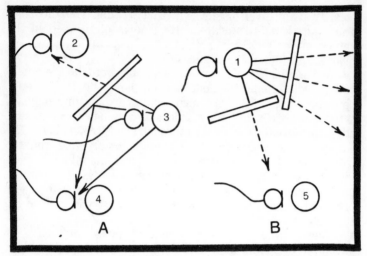

Fig. 2-9. Baffles can increase the separation between tracks.

energy to be troublesome. This factor is illustrated in Fig. 2-10. Whether or not baffle B shields microphone 2 sufficiently from the strong drum sound may be determined by the distance to and reflectivity of nearby room surfaces.

In Fig. 2-10 the two extremes are shown. A close, hard wall may direct enough drum sound to microphone 2 over a short path to completely nullify the effect of baffle B while a more distant soft wall would not. In the initial placement of baffles or the adjustment of their positions to correct poor track separation, many such possibilities must continually be borne in mind. We are beginning to see why good track separation demands plenty of space and, in general, walls and other surfaces with high absorption.

Frequency Effect of Baffles

It would be convenient if baffles were equally effective throughout the audible spectrum, but they are far from it. The physics of the situation tells us that a baffle is an effective barrier if it is large in terms of the wavelength of the sound being considered. As wave phenomena go, the audio band of ten octaves is an extremely wide band, yet that is what 20 to 20,000 Hz represents. The wavelength of sound at 20 Hz is about 56 feet; at 20 kHz it is but two-thirds of an inch! A baffle 4 feet wide is a good barrier at the high end, but at the low end the sound sweeps by like an ocean wave washing past a dock piling. At about 300 Hz this baffle is one wavelength wide.

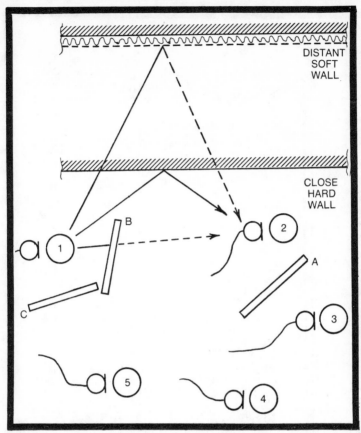

Fig. 2-10. The presence of nearby reflective surfaces can materially reduce separation between tracks.

Sounds at the bass frequencies go by as though nothing is there, *even if the baffle is made of lead a foot thick.*

Being aware of this basic limitation of baffles helps us use them intelligently and not expect them to do something contrary to nature. This deficiency of baffles causes us to look elsewhere for a solution to some of the problems of separation between tracks in our multitrack recording studio. The various nooks and crannies used in some studios to tuck away certain problem instruments can be considered an extension of the barrier principle into partially enfolding walls. Isolation booths can be considered an ultimate extension of the same principle by completely enclosing the source. This changes the problem from one of diffraction around a barrier to sound transmission through wall barriers.

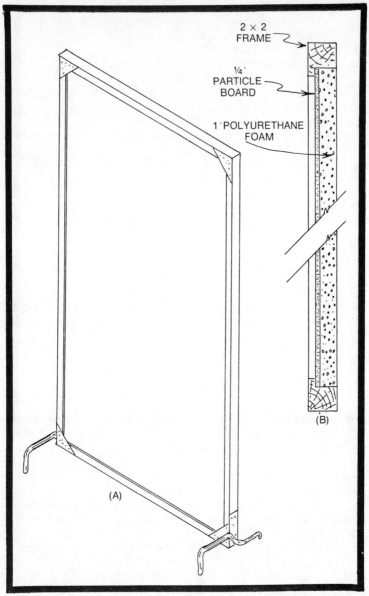

Fig. 2-11. Sketch of simple yet effective baffle.

Relying upon distance to achieve good separation within a musical group may cause other problems. Visual contact between conductor and musicians may be degraded, or one musician may not be able to hear the others well enough. For a large musical group, a studio may not be large enough to allow the needed separation. If one had the steady job of recording a symphony orchestra in a single location, the fixed-barrier arrangement of Fig. 2-1 might offer some promise. The average recording studio, however, must be prepared to jump from large groups to small, instrumental to vocal, traditional to popular—and usually at a moment's notice. This requires flexibility, and the portable baffle (Fig. 2-11) is the presently accepted method of achieving that flexibility. There may be recurring elements between diverse recording sessions which might encourage a degree of fixed-barrier facilities. For example, separate booths for drums and soloist and alcoves for partial separation of musical groups are often a fixed part of the studio scene. Beyond this, however, separation is generally achieved through the use of portable baffles.

Construction of Baffles

Acoustically, we want a baffle to do one or more of the following things for us:

- Serve as a barrier providing appreciable transmission loss between a high-level source and a nearby microphone of another track.
- Act as a reflecting surface to return sound energy to the ears of a musician or possibly to a microphone.
- Absorb sound

All three functions can be built into a single baffle (Fig. 2-11) by having one hard side and one soft side, such as plywood or particle board, which also offers significant transmission loss.

There are many materials which offer sufficient transmission loss to be suitable for use as baffles. We think first of boards, several of which are listed in Table 2-1 with estimates of the transmission loss they offer at different frequencies.

In general, the greater the mass of the barrier material, the greater the attenuation of sound as it passes through the barrier. For convenience, it is common to use *surface density*, which is the weight of a square foot of the material. For example, ¼ in. plywood weighing 0.75 lb/sq ft offers 10 dB transmission loss at 125 Hz, increasing with frequency to 35 dB at 4 kHz. As sound in the 125−250 Hz range tends to be diffracted around the screen no matter how thick or dense it is, the low transmission-loss figures at these frequencies tend to be immaterial.

Table 2-1. Transmission Loss (in dB)
of Some Baffle Materials

Material	Frequency, Hz						Data Source
	125	250	500	1000	2000	4000	
BOARDS							
Plywood, ¼ in. (0.75 lb/sq ft)	10	15	20	25	30	35	(Est by mass law)
Plywood, ¾ in. (1.9 lb/sq ft)	17	23	27	32	36	40	"
Particle board, ⅝ in. (2.5 lb/sq ft)	19	24	28	33	37	42	"
FABRICS							
Lead-loaded vinyl (0.5 lb/sq ft)	5	9	15	21	27	33	Ref. 6
Lead-loaded vinyl (1.0 lb/sq ft)	15	17	21	28	33	37	Ref. 6
TRANSPARENT SHEET							
Flexible PVC (70 mils thick)	11	12	15	20	26	32	Ref.7

Further, the ear's greatest sensitivity is in the 2 to 3 kHz region, not at the low frequencies. We conclude that 20 to 35 dB in the important region is a respectable transmission loss for the barrier to offer and, because of its reasonable weight, ¼ in. plywood may be considered a reasonable baffle material. Going to ¾ in. plywood increases the transmission loss to 27–40 dB, but the baffle would be almost three times heavier. Particle board is considerably denser and would make the baffle even heavier than ¾ in. plywood; the transmission loss would be only slightly better, 28 to 42 dB in the 500–4000 Hz range.

With the growing emphasis on noise control in industrial plants, many heavy fabrics are on the market for use as baffles around noisy operations such as punch presses, grinders, etc. to protect the hearing of nearby workers. In the studio, retractable curtains hanging from wires or tracks may offer some advantages as baffles. Such use is rare at the present time, except for the occasional use of ordinary fabric drapes, which are generally ineffective as sound barriers. Table 2-1 lists measured transmission losses on lead-loaded vinyl sheets of 0.5 and 1.0 lb/sq ft densities. These flexible sheets compete well with plywood.

Maintaining visual contact between members of a musical group is of greatest importance. Sometimes visual contact is desired between musicians who must be separated acoustically. One approach to this is a short baffle over which the musician can peer but which interposes a barrier for his instrument. Another is a taller baffle with a window in it. Heavy plate glass ports may be put in any ¼ in. baffle with little degradation as a sound barrier. In fact, short baffles may be topped with glass or plastic panels which offer the same transmission loss as opaque materials of the same surface density. These would be reflective on both sides, it must be remembered.

Flexible, transparent polyvinyl chloride (PVC) sheet only 70 mils thick is only 3 to 5 dB inferior to ¼ in. plywood in the critical range above 500 Hz. This material can be sewed into the 0.5 lb/sq ft lead-loaded vinyl sheet with no deterioration as a sound barrier. In fact, the hanging baffle could be made entirely transparent. So we see that transparent sound barrier materials are also available in the search for adequate acoustical separation.

One side of our baffle should be made absorptive if it is to have the greatest flexibility in the studio. Almost any kind of porous acoustic tile or sheet can supply the necessary absorption, but we also ask that this material be strong enough to withstand normal handling without abrasion. One approach is to cover softer materials with fabric. Another is to use materials with bonded facings to give the necessary wearing qualities.

The sound absorbing materials applied to one face of our baffle will not help much in attenuating sound passing through the baffle; we should select it strictly on the basis of measured sound absorbing qualities. Table 2-2 lists several typical materials suitable for the soft side of our baffle. An absorption

Table 2-2. Sound Absorption Coefficients of Absorptive Baffle Facings

Material	Frequency, Hz						Data Source
	125	250	500	1000	2000	4000	
Foam, polyurethane (1 in. thick)	0.13	0.22	0.68	1.00	0.92	0.97	Ref. 8
Foam, polyurethane (1 in. thick with perf PVC facing)	0.13	0.12	0.72	0.98	0.91	0.91	Ref. 9
Perf cellulose fiber tile (¾ × 12 × 12 in.)	0.09	0.26	0.81	0.99	0.81	0.51	Ref. 10
Glass fiber board glass cloth faced (1 in. thick)	0.03	0.17	0.63	0.87	0.96	0.96	Ref. 11

coefficient of 0.68 means that, under the standard conditions of the measurement, 68% of the impinging sound at that frequency was absorbed by that material. The listed materials run the gamut from polyurethane, faced and unfaced, through the common type of acoustic tile with holes or slits cut in the face, to glass fiber board with attractive glass cloth facing. With regard to the many foams on the market, be warned that there are, acoustically, two types. In one type the interconnecting walls of the tiny bubbles have been removed, creating an acoustically transparent foam used even for fancy loudspeaker grilles. In the other type the bubbly walls are intact and are effective in trapping and absorbing sound.

We have, then, suitable materials of which a portable baffle having a reflecting side and an absorbing side can be made. These baffles offer reasonable transmission loss to sounds above 500 Hz, even if provided with a window. Figure 2-11 shows one possible construction of such a baffle which is reasonably lightweight, relatively inexpensive, and effective. Baffles of full and half height complement each other. Brown[1] has devised a system of baffles based on 1 meter (3.3 ft) width and 1.3 and 2.3 meter (4.3 and 7.5 ft) heights which, with portable roofs, can be assembled into covered booths 2 meters wide and any integral modular length.

Cylindrical Baffles

The Japan Victor Company is making extensive use of stacks of semicylindrical diffusers as baffles (Fig. 2-12). Although bulky, these segments of cylinders have a unique contribution to make. The convex surface is commonly formed by wrapping a thin panel such as ¼ in. plywood around shaped ribs and fastening them so that they will not rattle when vibrating. Some of the sound impinging on these convex surfaces is absorbed and most of the rest is reradiated through a wide angle in the process. A flat panel of similar dimensions and material would be quite directional. Volkmann[2] points out that convex surfaces can direct sound energy toward musicians rather than toward the microphone, creating a livelier acoustic for the benefit of the musician in contrast to the overall dead characteristic of the studio.

General constructional features of such convex baffles are apparent in Fig. 2-12. Stacked three high as they are in this photograph, it is well to incline the top baffles downward, much as Volkmann did to his studio walls. This tends to direct downward the energy reradiated from the cylindrical surface. For most flexible use in multitrack studios, the backs of the baffles should be absorptive. Filling the cavity with glass fiber

Fig. 2-12. Convex baffles used in the studios of Japan Victor Company in Tokyo. These serve not only as shields between instrument sections, but to return sound to the musician's ears so they can hear themselves.

should accomplish this, holding it in place by a wire mesh which could be covered with fabric for a more attractive appearance.

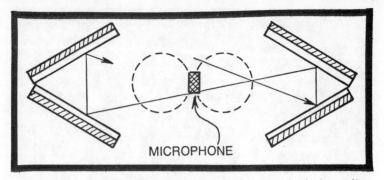

MICROPHONE

Fig. 2-13. Vee baffles of the type used extensively in radio dramatics to simulate outdoor acoustics.

Vee Baffles

The art—or more accurately, science—of using baffles to achieve track separation is certainly not new, as we saw in Disney's *Fantasia*. Nor is their use new with respect to dramatic effects, for they have long been used in radio to achieve special effects, especially the effect of being outdoors.[5] Open-air acoustics are distinguished by almost complete lack of reverberation. To represent such a condition indoors, the anechoic chamber comes to mind, but such extremely dead rooms have been proved to be quite impractical and compromise techniques have been adopted.

One of the most widely employed substitutes is the double-vee arrangement used commonly with a ribbon microphone as shown in Fig. 2-13. The sound picked up by the microphone is dominated by that which is confined to the space between the two baffles, as shown. The pickup of studio reverberation is minimized as the nulls of the microphone pattern are pointed toward the openings between the baffles. Because of the short path between the vee elements, the number of successive reflections is very high, causing the sound to decay rapidly, hence the "outdoor" characteristic.

Serious standing-wave colorations of the sound result if the vees are set at too large an angle so that the faces are parallel to each other. If the soft sides of the baffles are placed toward the microphone, the high frequency components of the sound will be reduced. With the hard sides toward the microphone, the highs will be enhanced. Baffles of any practical dimensions are, as we have seen, ineffective at the lower frequencies. Those in the double-vee form certainly do not duplicate outdoor acoustics, but the sound is sufficiently different from the normal studio sound to serve as a substitute which is often acceptable. Whether this arrangement of baffles might be useful in the multitrack studio remains to be seen.

Umbrella Baffle

Large beach and patio umbrellas of heavy material are being used to provide a modest amount of overhead baffling. Most commonly used over the drums, such an umbrella could help somewhat in obtaining a tighter sound and in reducing leakage to other tracks, especially in the presence of a low, reflective ceiling.

SEPARATION AND STUDIO ACOUSTICS

Anyone who has worked with microphones is aware of the fact that placing a microphone close to a source minimizes room effect in the pickup. In an anechoic chamber or outdoors away from reflecting surfaces, the sound from a point *source* is radiated equally in all directions and the wavefront is said to be spherical. Under such conditions the sound level falls off 6 dB for each doubling of the distance from the source. This "outdoor" condition is approached close to a source, even inside a reverberant studio. As the distance from the source is increased, however, a point is reached where reverberation prevails. The sound level tends to remain constant from that point out. The deader the studio, the greater the distance from the source the spherical divergence situation prevails.

All of this tells us that the deader the studio acoustically, the more freedom we have in microphone placement to minimize the effect of room reverberation. And the very basis of separation recording is to record the separate tracks as dry as possible to retain maximum freedom of track manipulation later. If we record a very dry track, artificial reverberation can be added later in any desired amount. If room reverberation is a prominent component of the recorded track, there is no way to remove it and some of the mixdown options are lost. For these reasons, the general practice has developed of building quite dead multitrack studios. The walls of such studios have low average reflectivity which helps solve the microphone placement problem illustrated in Fig. 2-10.

ELECTRONIC SEPARATION

Electronic noise gating circuits can be used to achieve better track separation during recording or for the correction of spillover from one track to another during mixdown. Gating action is obtained by a threshold adjustment by which signals below the threshold level are rejected, those above are passed as shown in Fig. 2-14. Let us assume that the signal on track 3

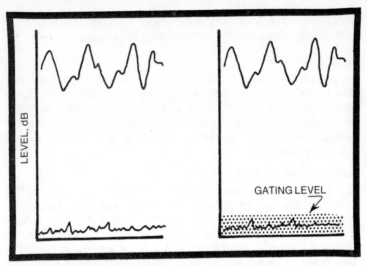

Fig. 2-14. Separation may be aided by the use of noise gating circuits which eliminate low-level signals from adjacent sources.

is spreading onto tracks 5 and 8 too much and we wish to clean up tracks 5 and 8. The spreading sound, even though quite audible on tracks 5 and 8, is actually separated from the desired signal by a significant difference in level. The threshold of a noise gate in each of channels 5 and 8 could be increased just enough to eliminate the interfering signal with negligible effect on the higher level desired signal.

CONTACT PICKUPS FOR SEPARATION

Electrically amplified instruments depend on special transducers for the input signal to their amplifiers. The transducers are attached directly to the instruments. The pickup for the violin, viola, cello, or string bass is built into or attached to the bridge of the instrument. For brass instruments it is mounted in a hole drilled in the mouthpiece. In guitars and other fretted instruments the transducer is attached to the tailstock with a special adhesive. Transducers are available for drums, harps, pianos, flutes, and piccolos, among others.[12] Practically any instrument can be made into an amplified instrument with such a transducer and an appropriate preamplifier, amplifier, and loudspeaker.

Although the usual justification for using such pickups is greater volume, a tremendous advantage is the fact that the electrical signal can be fed directly into an input channel of a console. In so doing we have replaced a microphone sensitive to airborne sounds with a transducer sensitive only to instrument vibrations.

A musical group with each instrument's transducer fed directly to the mixer has a very great separation between tracks. The musicians can even talk to each other during the recording without audibly affecting the tracks. Such an obvious and complete solution to all studio acoustic and separation problems has not caught on extensively, probably because of quality factors, but it does offer a means of getting great intertrack separation.

REFERENCES

1. Brown, Sandy. *Recording Studios for Popular Music*. Proc. 5th International Congress on Acoustics. Liege. 1965 (paper G-36).

2. Volkmann, John. *Acoustic Requirements of Stereo Recording Studios*. Jour. Audio Engr. Soc., Vol. 14, No. 4. October 1966 (pp 324–327).

3. Rettinger, Michael. *On The Acoustics of Multitrack Recording Studios*. Jour. Audio Engr. Soc., Vol. 19, No. 8. September 1971 (pp 651–655).

4. The author is indebted to Erwin L. Verity, Production Manager, Walt Disney Productions, for his tenacity in running down old records and photographs concerning the production of *Fantasia*. Thanks also to Michael Rettinger for his private communications. He was one of the RCA engineers engaged in work on *Fantasia* for Disney and who, along with John Volkmann, designed the wheel-shaped structure of Fig. 2-1.

5. Nisbett, Alec. *The Technique of the Sound Studio*. Hastings House, New York. 2nd Ed., 1970 (Chapters 5 & 6).

6. Coustifab Curtains, manufactured by Consolidated Kinetics. 249 Fornof Lane, Columbus, Ohio 43207.

7. Ferro Coustiview 5, distributed by Consolidated Kinetics.

8. Ferro Coustifoam 3-D, distributed by Consolidated Kinetics.

9. Ferro Cousti-Headliner, distributed by Consolidated Kinetics.

10. PyRotect Standard Drilled, manufactured by Simpson Timber Company. (Most perforated acoustic tile has similar characteristics; this type of facing will not stand abuse.)

11. Owens Corning Fiberglas Corporation's linear glass cloth board. Architectural Products Division, Fiberglas Tower, Toledo, Ohio 43659.

12. The Barcus-Berry line of transducers is distributed solely through the Chicago Musical Instrument Co., 7373 N. Cicero Avenue, Chicago, Illinois 60646.

3 The Audio Mixing Console

Multitrack recording requires complex equipment to give the needed flexibility of control. Further, the skill of the operator must match this complexity if the full potential of the equipment is to be realized in practice. Mixing consoles studded with several hundred knobs and controls and with a dozen or so VU meters are very common. However, a knob controlling a given function may be repeated for every channel, so things are not as complicated as a novice's first impression might indicate.

Ideally, the audio mixing console should be considered a creative tool. The operator needs musical insight, keen ears, and an analytical mind tempered by years of experience in sound. He must master every function of the board; handling of the controls must become instinctive and second nature to him as he concentrates on sound quality. He can do this job best with a good understanding of what goes on behind the knobs and buttons. Fortunately, the literature is rich in this subject,[1-15] but the information is fragmented and incomplete We shall try to pull it all together into a more logical and understandable form.

THE FUNCTIONS OF A CONSOLE

Let us first approach this complex dragon by considering the various functions it is expected to perform.

1. *Amplification* is provided to do these jobs, singly or in combination:
   - Bring weak microphone signals and higher level line inputs up to usable levels.
   - Provide isolation between critical circuits to avoid interaction.

- Compensate for losses in dividing, switching, and attenuating networks.
- Provide modest amounts of power for cue headphones, reverberation devices, and *talkback* loudspeakers.

2. *Controllable equalization* is available in each channel for these reasons:
   - Compensate for frequency deficiencies (peaks or excessive attenuation) of incoming signals.
   - Achieve a variety of dramatic effects.
   - Discriminate selectively against unwanted noise.
   - Provide maximum naturalness in the reproduction of voice.
3. *Channel assignment.* Switching facilities are provided to route any input channel to any output bus (major circuit).
4. *Aural monitoring* to evaluate not only the quality of signal in each channel, but the general quality of stereo or quadraphonic combinations.
5. *Visual monitoring* by VU metering to avoid overdriving electronic circuits or oversaturation of magnetic tape. The standard VU meter is still used extensively in this capacity, although banks of 16 or 24 of these do present something of a problem with respect to neck swiveling and plain comprehension.[12] Information on peak as well as average or RMS (root-mean-square) level is needed.
6. *Mixing* to allow selective combinatory control. In the major mixing system of the console the many input channels must be brought together in varying combinations and sent out on 1, 2, 4, 8, 16, or even more output buses. There may be submixing systems for monitoring or for cue (foldback).
7. *Fader.* A fader (potentiometer) in each channel makes possible the continuous adjustment of the level of the signal on that channel from off to normal level. The assembly of faders of channels assigned to a given output bus establishes the balance or relative contribution of each channel to the composite bus signal in mixdown.
8. *Reverberation.* Circuit switching and level adjustment facilities are provided so that reverberation may be added to any channel in any amount desired.
9. *Foldback.* In separation recording, musicians often have difficulty in hearing each other, or even hearing themselves. One approach to this problem is to equip them with headphones fed from the console. This

usually requires a submixing and amplifying system. Foldback is also required for overdubbing in which subsequent tracks are added to those originally recorded.

10. *Talkback*. To maintain verbal contact between console operator and performers in the studio, a special microphone, amplifier, and studio loudspeaker are required. The studio microphones enable the console operator in the control room to hear the performers on the monitor loudspeakers. When the console operator activates his talkback circuit, the other circuit must be disabled momentarily to prevent howling. The studio talkback loudspeaker may serve also for playback.

11. *Panoramic potentiometers*, commonly called *panpots*. In the simplest setup for stereo recording the panpot of a given channel allows the "moving" of that voice or instrument anywhere from the extreme left to the extreme right in the auditor's conscious field of hearing. Through similar pots the ambience in quadraphonic systems may be built up at will with signal contributions from appropriate channels.

12. *Test tones*, which are used for these test functions:
- Line up the various channels before a recording job.
- Check frequency response of each channel.

13. *Jack field*. Key points of the console electronics are routed through connectors in the jack field. The jacks are wired so that all standard functions are *normaled* through, requiring no patching at all. The jack field provides these capabilities:
- Facilities testing.
- Makes possible the introduction of outboard equipment without tampering with the console wiring.
- Increases the flexibility of the board by making it convenient, via plug-in patchcords, to reroute channels and change the functional configuration of the board.

The audio mixing console is the heart and brain of the multitrack recording system, serving the needs of input (music, speech, multitrack playback, etc.) and output (multitrack recorder, disc recorder, etc.). To do this well, certain ancillary equipment is usually involved. Such equipment may be mounted in a rack in the control room or at a remote location. Ancillary equipment might include:
- Speaker/monitor power amplifiers.
- Noise reduction equipment, which has signal-controlled frequency response to reduce subjective noise level.

- Limiters and compressors to increase the average level without overdriving, and make better use of limited dynamic range.
- Expanders to increase dynamic range and improve apparent signal-to-noise ratio.
- Reverberators. Although the echo send and echo return circuitry is a part of the console, the actual reverberator may be mounted in a rack in the control room, or as a separate unit in another room. If it is a reverberation chamber, it may be at some distance from the control room.

Input Module

If the overwhelming first impression given by a typical multichannel console is followed by a second, more analytical glance, it will be noticed that the host of knobs and switches are, in general, arranged in neat vertical rows and that identical rows are repeated. On the left portion of the board each row represents a separate input channel, each one of which, during a recording session, is fed by its own microphone picking up the signal from a vocalist, a single instrument, or a small group of instruments of similar type. During a session each channel remains associated with its single component part of the group musical effort and is probably appropriately labeled (DRUMS. BASS. PIANO. VOCAL. etc.) with a grease pencil on the strip provided.

Input Control Section

The microphone signal is first routed through the top of the input module as shown in Fig. 3-1. For maximum separation, each microphone must be close to its assigned source. Some musical instruments and some vocalists holding the microphone close to the tonsils, and certainly instruments with electrical amplification create such high microphone output that the preamplifier stage may be overdriven, causing distortion. To care for such situations, most consoles provide a switch by which a fixed amount of attenuation can conveniently be inserted between the microphone and the preamplifier. This may be 30 or 40 dB maximum, adjustable in 10 dB steps.

A variable microphone attenuator is also a part of the input control to enable the operator to place the channel slide fader on an appropriate part of the scale. This attenuator usually controls the gain of the preamplifier by varying the amount of feedback. The fixed and variable level controls allow the various input microphones to be preset to a uniform

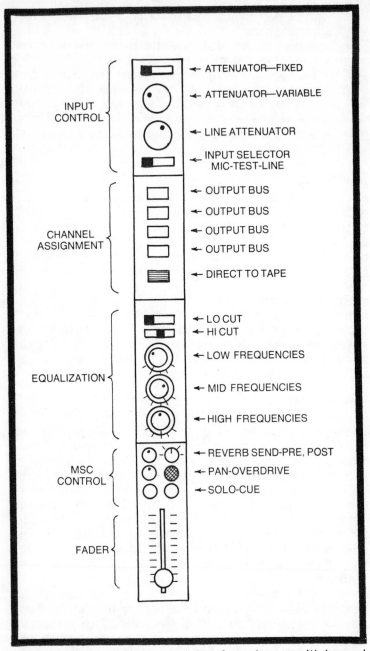

Fig. 3-1. Typical input module of modern multichannel audio console.

maximum output level, regardless of the output levels of the individual transducers involved.

The input control section of the input module probably also contains a selector switch so that the channel can be fed by signals from a microphone, a line, or a test oscillator. The microphone input handles low-level inputs, and the line input handles high-level inputs. Line attenuators allow fixed or variable control of the incoming audio signals. Connecting the test oscillator to the channel provides a reference signal useful in lining up the channels in preparation for a recording session and for calibrating the console VU meters. The frequency of the test oscillator may also be varied to check frequency response of the channel or to verify settings of the equalizer controls.

Channel Assignment Section

On the right side of the console are the program output buses. The console is selected to have 2, 4, 8, or 16 output buses, whatever is needed for the job at hand. Let us assume that our board has 16 input channels and 4 output buses. This will enable us, among other combinations, to mix down 16 microphone signals on 16 magnetic tracks to a 4-track master tape. This would also enable us, during a recording session, to monitor in mono, stereo, or quadraphonic.

In the interest of complete flexibility, we want to be able to switch any input channel to any output bus. Furthermore, we would like the privilege of delegating several input channels to a single bus which requires a cumulative or summing mode. This brings us to the channel assignment section of our typical input module of Fig. 3-1. The upper four buttons represent the four output buses. By pressing the top button, this input module is connected to output bus 1. If we also press button 2, this input module would also be connected to output bus 2. In a similar way buttons may be pressed on other input modules connecting them to any selected output bus. Pushing button 5 connects the output of this input module to the corresponding input of the multitrack recorder.

Equalizer Section

One of the big differences between the mixing consoles of a few years ago and modern desks is the profusion of equalizing facilities available today. Tremendous flexibility and compact size are the result of the application of active circuits involving solid-state devices which have almost completely replaced the passive circuits with their bulky inductors and capacitors. Each input module of the modern console—even those mixers used for sound reinforcement in better systems—has its own equalization to alter the frequency response of the channel.

Referring to the equalizing section of Fig. 3-1 we see one possible arrangement to achieve the wide range of adjustment demanded today. First, there will probably be a simple low-cut (high-pass) filter to chop off energy below a selectable frequency, say 40 Hz or 100 Hz. This filter attenuates unwanted rumbles such as those from air-conditioning equipment and room resonances, and guards against overdriving the amplifier. This filter might roll off at a 12 dB per octave rate to avoid undesirable effects of very steep filter skirts.

A high-cut (low-pass) filter may limit the high frequency edge of the passband. By a simple rotary or slide switch, the frequency at which this filter takes effect is adjustable to two or three frequencies such as 5, 10, or 15 kHz.

One of the ways the individuality of the many consoles on the market is expressed is in the configuration of the band equalizers. Two, three, or even four adjustable points can be provided between the low-cut and the high-cut filters. In general, the early tone-control idea of one adjustment of high frequency response and one for low frequency response has been augmented by the need for at least one *presence* equalizer in the midband region to improve speech intelligibility or accomplish other tailoring jobs. These equalization points provide 12 dB boost or dip in the selectable two-, three-, or four-frequency regions. Concentric controls are used to conserve panel space. For example, one control adjusts the frequency at which a given filter is effective; a second one, arranged concentrically, adjusts the amount of boost or dip in that frequency region. A remarkable refinement of response shaping is possible with a three-band equalizer. In fact, it takes a very adept operator to come close to exploiting the possibilities of such equalizers.

Now that we have considered the typical equalizing section of the input module of Fig. 3-1, it must be noted that *there is really no such thing!* Some of the real-life consoles on the market exceed the degree of flexibility described here; others fall short of it. None matches it exactly. Table 3-1 summarizes the equalizer section capabilities of several of the consoles on the market today (selected at random).

Miscellaneous Control Section

In our hypothetical input module we will lump a number of necessary controls in a miscellaneous control section, close to the hand of the operator resting on the channel fader. Two of

Table 3-1. Equalizer Capabilities Console

Mfr. Model	Band 1	Band 2	Band 3	Band 4	Remarks
Audio Designs 770	40,80,160 Hz stepped ±12 dB	200,350, 560,900 Hz stepped ±12dB	1.5, 2.5, 4.3, 7.2 kHz stepped ±12 dB	10, 12.5, 15 kHz stepped ±12 dB	
Audiotronics 501	80,150 Hz cont var ±12 dB	300,600 Hz cont var ±12 dB	1.6, 4 kHz cont var ±12 dB	7.5, 12 kHz cont var ±12 dB	Low-cut High-cut switches
International Telecomm. Inc. MEP-130	LF shelf cont var ±12 dB 5 dB/octave	100 – 8000 Hz cont var Q variable 4 – 14 dB/octave	400 Hz – 24 kHz cont var Q variable 4 – 14 dB/octave	HF shelf cont var ±12 dB 5 dB/octave	
MCI JH-416	2 freq sel by button. 11 pos shelf ±12 dB	12 freq sel by switch. 8 pos peak and boost-cut switch ±14 dB	2 freq sel by button 11 pos shelf ±12 dB		
Neve 1081 Amplifier	330,180,100, 56,33 Hz shelf or peak, cont var ±18 dB	1200,1000, 820,680, 560,470, 390,330, 270,220 Hz cont var ±12 dB or ±18 dB	8.2,6.8,5.6, 4.7,3.9,3.3, 2.7,2.2,1.8, 1.5 kHz cont var ±12 or ±18 dB	3.3, 4.7, 6.8, 10, 15 kHz shelf or peak, cont var ±18 dB	Low-cut: 27,47, 82,150,270 Hz 18 dB/octave High-cut: 18,12, 8.2,5.6,3.9 kHz stepped, 18 dB/octave
Neve 1084 & 1085 Amplifier	220,110, 60,35 Hz cont var ±18 dB (1085-stepped)	7.2,4.8 3.2,1.6, 0.7,0.35 kHz cont var ±12 dB or ±18 dB (1085-stepped)	10,12,16 kHz shelf cont var ±18 dB (1085-stepped)		Low-cut: 45,70, 160,300 Hz High-cut: 18,14, 10,8,6 kHz 18 dB/octave stepped
Spectra-Sonics 1024-24/8 (502 Equal.)	50,100,200, 300,400 Hz ±12 dB stepped	500,800 Hz 1.2,1.6,2.0 kHz ±12 dB stepped	2.5,3.5,5.5, 7.5,10 kHz ±12 dB stepped		LF Shelf: 50 Hz HF Shelf: 10 kHz
Tascam 10	90,200 Hz cont var ±10 dB	3,5 kHz cont var ±10 dB	10 kHz cont var ±10 dB		Low-cut: 40,100 Hz High-cut: 5,10 kHz 12 dB/octave

Abbreviations: LF = low frequency   cont var = continuously variable
                HF = high frequency   sel = selectable
                freq = frequency

these controls are dedicated to reverberation. The signal of input module 1 may be sent to any one of the four reverberation chambers, or reverberators, either before or after the channel fader. This is done by selecting the PRE or POST point on the selector switch. The other knob labeled REVERB SEND is a potentiometer by which the magnitude of the signal sent to the reverberator may be adjusted. Some reverberators, notably those of the spring type, are quite susceptible to overdriving, hence the desirability of being able to control the send level.

It is unfortunate that the word *echo* is widely used instead of *reverb*. Echo is a return of a discrete bundle of sound energy after a given time, such as an echo from a highly reflective wall; reverberation is a statistical decay of sound energy in a room over a finite period of time. Reverberation enhances music and gives it an expansive quality. Except for special effects, real echo is not a desirable effect.

Below the reverb controls of Fig. 3-1 is an indicator that lights up when the signal of the channel goes into the nonlinear region of the associated amplifier's response characteristic. The operator is warned by this lamp when transient peaks (which the VU meter is too slow to follow) overdrive amplifiers, causing distortion.

Alongside the overdrive indicator is the panpot. When more than one channel is assigned, this PAN control is automatically connected. *Rotating* this control counterclockwise shifts the sounds of this channel to the monitor loudspeaker covering the left output buses. Rotating it to the right shifts the sound of this channel to the right.

By pushing the SOLO button, this channel and this channel alone is connected to the monitor loudspeaker.

The CUE knob controls the volume of the signal from this channel sent to the cue bus. For example, earphones driven from this cue bus can be worn by musicians in the studio so they can hear appropriate musical cues from other members of the recording group. This is also the means by which synchronism is obtained when postrecording another channel (overdub) to be mixed with the original recorded tracks.

Fader Section

The fader controls the level of the input module signal which is sent on to the output bus. The linear, stepless attenuator has almost universally replaced the rotary stepped fader of past times. The linear faders are built around resistance elements having long life, a good feel, and noiseless operation.

We have described a typical input module which probably differs in some respects from all real-life input modules on the market, yet bears similarities to all. This approach has allowed us to concentrate on the functions without the complications of working around proprietary features which may be excellent operationally but which complicate our understanding of the basic functions.

We can mount the input modules side by side and have a console of sorts. We could feed each with a microphone line from the studio and connect the 16 outputs to the inputs of a multitrack magnetic tape recorder and plunge immediately into recording. But this would be an extremely inflexible arrangement without even monitoring facilities, to say nothing of talkback, cue, reverberation, and all the other little pushbutton editing conveniences which contribute so much to overall efficiency. Most recording engineers would probably agree that a multichannel recording console really needs mixing facilities to do a good recording job. They need to hear something close to the final mixdown to make good judgments while laying down the tracks on the tape. This doesn't commit the producer to follow the same relative levels between performers, channel equalization, or that particular blend of reverberation; but it does give some idea of how the band sounds when put back together, even on a temporary basis, after having been dismembered by separation recording techniques.

Once we have provided a mixing facility to meet the recording need, we have a console which can function both for multitrack recording and later as a mixdown board. Its use in mixdown might run into conflict with recording schedules, but operationally the same board can do both jobs well. Consoles to be used solely for mixdown can be somewhat simplified, as we see in Chapter 9.

Output Bus Master Module

Let us now consider the additional units needed on the right side of the board to help us in using the 16 input modules on the left, each of which is patterned after Fig. 3-1. Each output bus must have its own fader (Fig. 3-2). The 16 input modules are connected to these output lines by pressing the channel assignment buttons on each module. Any desired pattern of input channel delegation may be programed using these buttons. Summing circuits combine the signals in any combination without noticeable interaction. All four output faders are used for quadraphonic mix, two for stereo mix, and any one for mono.

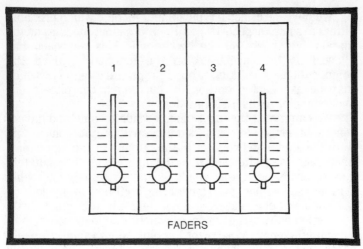

Fig. 3-2. Typical output bus master fader module.

Master Fader Module

Once the master output bus faders are set for mixdown, it is not convenient to fade the overall program in or out by opening or closing all four faders simultaneously. This is too much of a gymnastic exercise, and it disturbs the settings. An overall master attenuator, shown in Fig. 3-3, controlling four ganged faders, serves this purpose much better. It is traditionally placed to the right of the output bus fader.

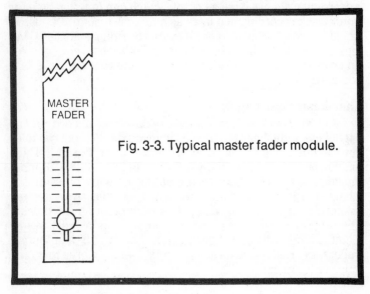

Fig. 3-3. Typical master fader module.

Cue (Foldback) Module

On the input module of Fig. 3-1 there is a pushbutton labeled CUE just above the channel fader. If this button is pressed on, say, channel 4, the channel 4 output is then connected to rotary attenuator 4 on the cue module of Fig. 3-4.

An example will illustrate one use of the cue facility, often called foldback. During a recording session, musician A has a couple of phrases of very tight rhythm with musician B who has the lead but who is so well separated in the studio that A has trouble hearing him. Pushing the CUE button on the lead musician's input module, say 12, feeds his signal to the cue bus through attenuator 12 on the cue module. Through a pair of headphones fed from this cue bus, musician A is provided with foldback sound from his lead, enabling him to do his stuff with distinction. Another use of this cue facility is in overdub. Let us say that musician A wasn't even present during the recording session and that his track on the tape is blank. The tape is played back with appropriate channels feeding the cue bus to A's headphones. Musician A can now close his eyes, imagine the boys are all in their accustomed places with him, and record his empty track just in time for the mixdown. We can see that this cue feature can be very valuable: in fact, fancier boards may have two or more cue buses so that different cue mixes meeting different needs can be fed at the same time.

Monitor Module

Figure 3-5 shows a possible arrangement of monitoring facilities. We are assuming that the control room is equipped with four loudspeakers arranged for quadraphonic monitoring or playback, each with its own power amplifier. For any given monitoring job it is then necessary to select the configuration of loudspeakers required for the job at hand. The monitor module of Fig. 3-5 has pushbuttons to select the quad, stereo, or mono mode. Another button allows the control room loudspeakers to play the signal on the cue bus. A dual pushbutton connects the monitoring loudspeakers to either the output bus or the tape playback. Loudspeaker level is controlled by four submaster attenuators followed by a four-gang monitor master level control.

Talkback Module

Talkback facilities are vital for relaying instructions and recording voice identification on magnetic tapes. The latter is called *slating*, a hangover from motion picture procedures in which a slate clapper with scene number, take number, and

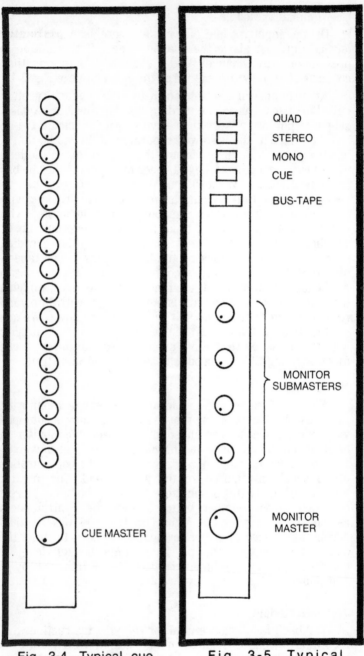

Fig. 3-4. Typical cue (foldback) module.

Fig. 3-5. Typical monitor module.

other production data is exposed on a few frames at the beginning of each take. In Fig. 3-6 is a talkback module having switching and level controls as well as the talkback microphone. The microphone may be mounted flush with the panel, intercom style, or on a *gooseneck* microphone mount.

Talkback routing buttons are of the push-to-talk type; rotary attenuators allow adjustment of level for studio talkback, communications with those receiving foldback, and slating of recorder tracks during both recording and mixdown. Circuit precautions must be taken to avoid howling while talking back to the studio in which there are open microphones.

Reverberation Module

Each input module (Fig. 3-1) has two reverberation controls, one selecting a prefader or postfader point for picking off the signal to send to the reverberator, and the other controlling the level of this signal. In the present case we are assuming that four reverberators are available; Fig. 3-7 shows the reverb module controlling the reverb return. The reverb assign selector switches make possible the connection of the return from any of the four reverberators to any of the four output buses in any pattern desired and without interaction. The level of the return from each of the reverberators is

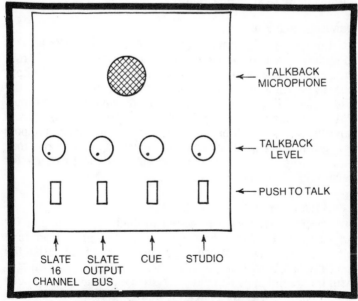

Fig. 3-6. Typical talkback module.

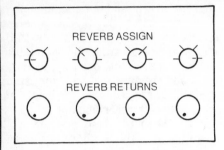

Fig. 3-7. Typical reverberation (echo) module.

adjustable by the rotary attenuator under the assignment selector switch for that reverberator. To achieve maximum flexibility, several reverberators are required because different instruments and different special effects require different reverberation times. It is far more than a matter of simply restoring reverberation lost in the use of a dead studio for separation recording. One may wonder about feeding many channels into a single reverberator. Why be concerned? In real-life many instruments in a band feed signals into a single reverberator, the concert hall, with no chance of adjusting anything. Our board and some separation between tracks is far, far more flexible than that.

Quadraphonic Panning

For genuine quadraphonic *effect*, it is necessary to be able to move any sound source freely from right to left, from front to back. After delegating our input modules to the four output buses according to some tentative plan of placement in the auditory field, it is now necessary to control each output bus. With the quadraphonic panpot a source may be moved anywhere in the auditory field. In Fig. 3-8 four such panpots are given the task of placing the sound signals of the four outputs wherever the producer desires in the final master recording.

Level Indicators

Controls have been provided in each input channel for adjusting the level of microphone signals or signals from tape playback. We now must provide some means of indicating the absolute levels of these signals. The VU (volume units) meter with its special ballistic characteristics has traditionally been the standard device for indicating audio signal levels. A row of these VU meters mounted above the main board near eye level

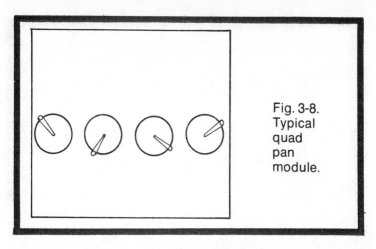

Fig. 3-8.
Typical quad pan module.

could be considered a sort of outboard module. A VU meter for each of the 16 input channels plus one for each of the 4 output lines would provide all the level indication we need and possibly more. There are several possible ways to reduce the number of VU meters to economize and to simplify the visual problem of scanning so many in a meaningful way. For example, the overdrive indicator on each input channel module tells when the signal level excursion reaches the distortion point. This warning that the top extreme of our dynamic range is being approached could be considered sufficient level information for certain types of signals. By providing suitable switching facilities it is also possible to reduce the number of VU meters required by switchable sharing between input and output.

In Fig. 3-9 the space above the quad panpots is vacant—just the place to lay the script or score. Or the patch panel could be located here. The Germans have a nice approach to the patch panel location problem: they recess the panel and mount a hinged transparent plastic cover over it. This keeps unauthorized fingers from disturbing the patch cords, still leaving room for the score or script.

Another point should be made about jack fields. The old Western Electric double plug is now replaced by single tip/ring/sleeve plugs and jacks in modern consoles, which saves much space and eliminates the old bugaboo of having a double plug reversed. But the jack field approach is a very important feature. If the board is used in the intended ways, all circuits are normaled through. If you want to do something a bit off the beaten path, it is most convenient to do so. Jack fields can also be used to locate trouble. Electronic switching

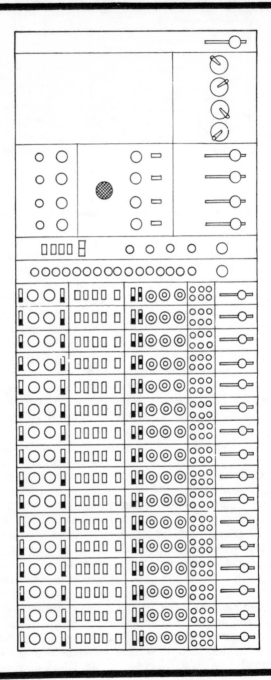

Fig. 3-9. Assemblage of the typical modules to make a multichannel audio console.

may eventually replace the mechanical jack field, but until that happens we are covered.

The console we have just assembled on paper is suitable for both recording and mixdown, but the mixdown feature is fully justified for recording because of the need for realistic monitoring and for playback, overdubbing, etc. This board is rather stark in its purpose of helping us see that a console with all those knobs and buttons isn't really so complicated after all. And every knob, button, and fader has a function, and the form of the board is determined by function—a good principle in multichannel/multitrack recording as well as in architecture. Let us now look at a few commercial audio control consoles.

There are strong similarities between the Buick, the Ford, the Chrysler, and all other automobiles; they have internal combustion engines, four wheels, rubber tires, a steering wheel, etc. But there are also differences which the salesman will be happy to share with you. These distinguishing features affect the cost, convenience of operation, as well as the overall quality of the finished product. Audio control consoles will inevitably bear a strong resemblance to each other because they are designed to do similar jobs. Similarly, each commercial console has its special features affecting cost and convenience, and each has its reputation for quality of components and workmanship. We cannot fully cover the features of any of the commercially available consoles in this chapter, but it is important to look briefly at a few representative boards. We shall see that each has its own way of arranging the basic functions we have put into our hypothetical mixer.

SPECTRA-SONICS MODEL 1024-24 CONSOLE

In Fig. 3-10 is pictured the Spectra-Sonics 1024-24 audio control system. We shall look rather carefully at this console to see what each knob, switch, and slider does. No other board is identical to this one, but all bear certain similarities. Despite the extensive variety of combinations and routing available to the operator, the control deck is laid out so that signal flow routing controls are logical, sequentially located, and readily mastered. This console features modular construction, as others do, with emphasis on flexibility and compactness. Each module is 1.5 in. wide; 24 inputs and 24 direct outputs are housed in a board only 60 in. long and 40 in. front to back. The modular construction lends itself to starting with fewer channels and expanding to full capacity later.

Fig. 3-10. Spectra-Sonics audio control console, Model 1024-24/8.

Input Module

Each channel is made up of three submodules arranged end to end with an assignment switching submodule at the top, input submodule in the middle, and the straight-line attenuator closest to the operator. Let us take these one at a time for detailed observation, beginning with the middle unit (Fig. 3-11) because the microphone or other input signal first enters here.

Line-Microphone Selector, Input Attenuator. The left toggle switch is moved to L for a higher level line signal, such as a tape playback, and down to M for lower level microphone signals. The right two toggle switches allow fixed attenuation of 10, 20, or 30 dB to be applied to either the line or the microphone circuit. This allows the channel signal level to be adjusted to a convenient part of the slide attenuator scale and provides protection against overdriving the preamplifier with very high-volume input signals.

Equalizer. This microphone/program equalizer has three pairs of knobs, one set to control the low frequency end of the spectrum (lower), one set to control the midfrequency range (center), and the final pair to control the high frequency range (upper). The three toggle switches at the bottom of this module allow the low (L), middle (M), or high(H) equalizer sections to be removed from the circuit if desired. The

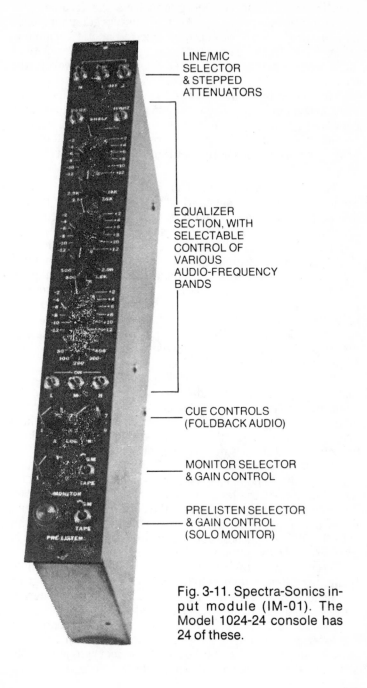

LINE/MIC SELECTOR & STEPPED ATTENUATORS

EQUALIZER SECTION, WITH SELECTABLE CONTROL OF VARIOUS AUDIO-FREQUENCY BANDS

CUE CONTROLS (FOLDBACK AUDIO)

MONITOR SELECTOR & GAIN CONTROL

PRELISTEN SELECTOR & GAIN CONTROL (SOLO MONITOR)

Fig. 3-11. Spectra-Sonics input module (IM-01). The Model 1024-24 console has 24 of these.

smaller, lower knob of each pair selects the frequency at which that section is to be effective; the larger upper knob adjusts the degree of equalization inserted and is variable in steps of 2 dB from −12 to 12 dB. The specific switch positions allow accurate logging for later use. In other words, the operator can boost or cut the level at each of 15 narrow bands of frequencies by 12 dB. Figure 3-12 shows the shape of the boost and cut curves at each of the 15 positions. Because these curves are plotted on an expanded scale, the skirts appear to be quite steep; but actually they are very gradual: 6 dB per octave, which is considered optimum by some for response shaping.

There are two toggle switches at the top of the equalizer section labeled SHELF, one marked 50 Hz and the other 10 kHz. The 50 Hz is at the low end of the equalizer low frequency spectrum and the 10 kHz is at the high end of the high frequency spectrum. If the left switch is set to a 50 Hz shelf and the lower selector switch knob is turned to 50 Hz, the shape of the 50 Hz curve of Fig. 3-12 is changed to give a flat shape in · the frequency range below 50 Hz. This is useful to eliminate rumble, low frequency room noises, or to "tighten up" the bass. If the 10 kHz shelf switch is actuated, the response will be flat from 10 kHz up, whether cut at any amount from −2 to −12 dB. This is most useful during mixdown to eliminate tape hiss from the tape machine tracks already recorded.

Cue. The two CUE knobs in Fig. 3-11 control the cue or foldback signals sent to headphones worn by musicians in the studio during recording to enable any one musician to hear any others, or during an overdub session when the other musicians are not present and a cue signal is needed for timing. Many consoles have the cue mixing controls gathered together on one or more cue panels on the right side. This board has two such controls, A and B, on each input module instead. The CUE A control adjusts the level of the signal on this channel. The CUE B control derives signal from the input module and would do the same for a separate cue mix (B) which could serve an entirely different purpose at the same time. (A third cue mix is available from the outputs of eight submix bus modules, as we see later.) Having the cue mix knobs on the channel module facilitates adding channels to the board as mix facilities grow proportionally with the number of channels without rendering obsolete a cue mix panel on the right end of the board.

Monitor. Even though the separate tracks are recorded at full level, unmixed, it is imperative that the mixer hears at least a tentative mix so that he can make accurate judgments affecting the finished mixdown product. There must be a

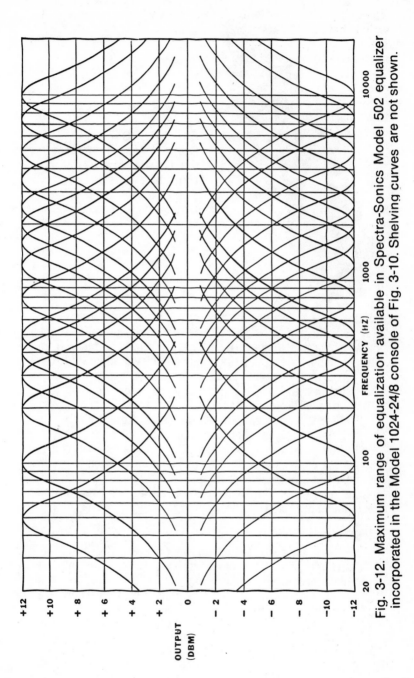

Fig. 3-12. Maximum range of equalization available in Spectra-Sonics Model 502 equalizer incorporated in the Model 1024-24/8 console of Fig. 3-10. Shelving curves are not shown.

69

Fig. 3-13. Spectra-Sonics' Model 901 straight-line attenuator. This is a stepless, continuously variable unit.

monitor mix which is for the control room only, having no effect on the tracks being recorded from the outputs. If the operator wants to hear the signal in this channel (from microphone or line input), the toggle switch would be set to the PGM (program) position. To hear the actual taped signal, it would be set to TAPE. The associated knob controls the contribution of this channel to the ultimate mix.

Prelisten. By pressing the PRELISTEN button, the mixer is able to listen to the signal of this channel alone on the monitor. This button overrides all other monitor sounds as long as it is engaged. It enables the operator to judge separation between channels or to locate and evaluate other types of problems, a channel at a time. The switch selects either the program on that channel or tape from the input of the module.

Attenuator

We shall now go to the attenuator module of Fig. 3-13. This is a straight line type of attenuator which has almost universally replaced the rotary attenuator in modern multichannel consoles used for recording purposes. There are many advantages to this type of attenuator, not the least of which is that the operator can see at a glance the setting of the various slide controls. Each such control attenuates the signal from 0 to full cut (∞), and is located next to the operator for obvious reasons.

Program Assignment

The third module of this channel is the switching module, Fig. 3-14. This module is the key to the overall flexibility of the console. The number of combinations which can be set up on the 19 switches and 2 controls of this module offer the creative operator almost unlimited freedom.

Submix. This board has eight submix lines; the pushbutton switches numbered 1 to 8 allow the operator to route the signal through this particular channel to any bus or to more than one bus in any desired combination by pressing appropriate buttons. The buttons light up when pushed, providing a constant indication of channel delegations. The bottom button, marked DIRECT, connects the channel directly to one input track of the associated multitrack recorder. This set of pushbuttons, then, provides an efficient way of routing the program of this channel in any desired way.

Quad Pan Control/Echo Send. For quadraphonic mixdown it is often necessary to move the signal of this and other channels to some specific position in the quadraphonic listening field. This is done with quad panpots. This control allows placement of this channel's signal left front (L_f), right front (R_f), left back (R_b), or at any position in between. The left one of the two controls is a dual rotary panpot control which is placed in the circuit by throwing the PAN/ECHO assignment

Fig. 3-14. Spectra-Sonics Model SM-02 input switching module.

switch to INPUT or MONITOR position. The INPUT position means that the signal comes from the input module, MONITOR means that the signal comes from the monitor remix control.

The right-hand knob is the ECHO SEND control. It adjusts the level of signal sent from this channel to the reverberation unit. The audio signal from this channel may be selected between INPUT and MONITOR by appropriate placement of the ECHO toggle switch.

Echo Select. The four toggle switches marked ECHO SELECT assign the audio to any one or all of the four echo lines (A. B. C. D). The operator may mix echo to the monitor without disturbing the input mix control setting. Thus, he may record without echo (dry) and monitor with echo (wet) to continually maintain an idea of what the finished product will sound like.

Monitor Select. There are four monitor outputs which normally feed the four power amplifiers driving the two front and two rear loudspeakers. The four MONITOR SELECT toggle switches allow the monitor mix of this channel to be routed to any selected monitor bus. Although marked A. B. C. D. the switches could as well be labeled L_f, R_f, L_b, and R_b if the four monitor outputs were connected to loudspeakers in that order.

The above description covers only one channel; the other 23 are identical. While the console has 24 inputs, it also has 24 outputs, as the DIRECT button on each channel connects directly to a track of the multitrack magnetic recorder(s) being used with the console. This is the arrangement for recording. For mixdown, up to 24 recorder tracks can be played into the line inputs of the 24 console channels and mixed down to one or all outputs (one for mono, two for stereo, or all four for quadraphonic), and sent to the master tape recorder. These outputs are located to the right of the 24 input channels along with two other modular strips. These right strips provide controls for such things as echo send master, echo return assignment, program submixing masters, quadraphonic masters, stereo and monaural masters, control room monitor assignment, studio monitor assignment, cue masters, signal generator, talkback assignment, and VU meter selection. We shall not go into these in the detail with which we have treated the input modules because the functions of many are apparent in the listing of them. Other equipment contained in the console includes talkback microphone for communicating with the studio and verbal identification of takes (slating), a 288-patchpoint jack field under a hinged door, remote recorder control buttons, etc.

A tremendous number of amplifiers are required in an audio control system of this kind. Modern solid-state

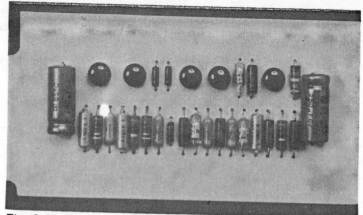

Fig. 3-15. Spectra-Sonics Model 110 audio amplifier. This is a solid-state plug-in amplifier used in audio control systems.

techniques have made possible high performance in a very small space. Normally all of the amplifiers of the console of Fig. 3-10 would be contained within the console itself except, of course, the monitor power amplifiers. The power supply with its hum generating transformer is also located external to the console.

The basic amplifier of the console we have been examining is shown in Fig. 3-15. This one printed circuit board, measuring only 2½ by 5 in., performs the function of microphone preamplifier where the signal is at the lowest level, and has a nominal 40 dB gain which can easily be increased to 55 dB if required. Only one other type of amplifier is used in the console at the less critical stages, which is an integrated circuit plug-in component. This allows on-site replacement for maximum serviceability in the field.

AUDIO DESIGNS NRC SERIES CONSOLE

An example of the NRC series of recording consoles offered by Audio Designs and Manufacturing, Inc., is shown in Fig. 3-16. In studying some of the features of this board we shall see further similarities in function and basic layout to other consoles, with notable differences in details of construction, circuitry, and subsystem arrangements. The console pictured can accommodate 24 input channels. Each input channel is composed of three modules: a linear-action attenuator, an input module, and a switching module. We shall consider these individually as they establish the basic character of the board.

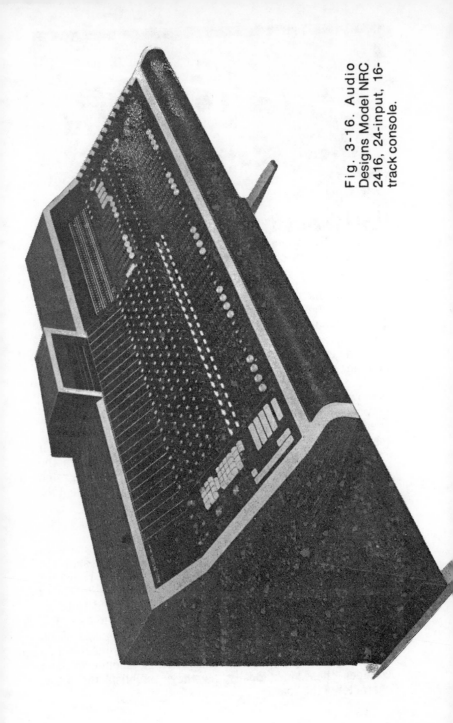

Fig. 3-16. Audio Designs Model NRC 2416, 24-input, 16-track console.

74

Fig. 3-17. Linear at-
tenuator featuring
spiral construction
(Audio Designs
Model 483 Slidex).

Linear Attenuator. The straight-line attenuator, Fig. 3-17, has a scale calibrated from 0 to infinity. Sliding the control causes a nylon roller wheel to move up or down, thus turning a spiral-cammed ribbon attached to a rotary control shaft. This arrangement insures precision of repeatability.

Switching. At the top of each channel is a switching unit utilizing magnet reed switches. Each module contains all the reed switches and solid-state circuitry required to route the signal of this channel to any of the program outputs or echo send channels. This module is actuated by a separate pushbutton control module, located on the left side of the board, which is shared with all other channels. Thus, all switching functions are accomplished by this single control module. This control module has three groups of pushbuttons; one group permits selection of the input channel to be delegated, a second group allows selection of the desired program bus, and a third group selects the echo channel. Individual buttons are provided for program clear and echo clear. A SOLO button is also provided so that any selected input may be heard through the monitor by pressing this button. One remaining button, marked CUML, permits selection of canceling or cumulative bus selection. In the canceling mode, an input may be delegated to a given output bus; should another bus button be pressed, the first selection will be canceled in favor of the second. In the cumulative, any number of audio lines may be selected simultaneously. A

Fig. 3-18. Audio
Designs Series 770
input module.

computer-type lighted digital readout on each channel's
switching module indicates the status of that channel.

Input Module. Between the straight-line attenuator next to
the operator and the switching unit farthest from the operator
is the input module (Fig. 3-18).

Mic/Line. The upper dual concentric control selects
microphone input or line input and allows insertion of up to 30

dB attenuation in 10 dB steps in either the microphone or line inputs. The L.L. and H.L. placards stand for low-level (microphone) and high-level (line).

Cue. The second control from the top, marked CUE, is another dual concentric type by which either of two cue circuits may be selected and the level of the signal controlled.

Equalizer. The four-band equalizer covers, from top to bottom, the following frequency bands; high frequency (HF), mid-high frequency (MH), mid-low frequency (ML), and low frequency (LF). Each control is also a dual concentric, one selecting the frequency at which the equalizer section is effective and the other the degree of boost or cut equalization. The great flexibility in the adjustment of channel frequency response by this equalizer is illustrated by the response curves of Fig. 3-19. Note especially the wide adjustment of low- and high-frequency shelving characteristics available as well as peaks and notches.

Echo. Below the four-band equalizer of Fig. 3-18 is the ECHO send control. This is also a dual concentric control, one control assigned as echo send level control and the other a three-way switch which injects the echo send signal before the channel attenuator (PRE), after the attenuator (POST), or after the equalizer (EQ).

Below the ECHO control are two paddle switches, one for equalizer (EQ) in and out, the other for echo on and off (ECHO).

Overload. A tiny pilot lamp comes on if any peak signal excursion (thus exceeding the dynamic signal handling capability of subsequent amplifiers) goes into the distortion zone.

Solo. Beneath the overload light and immediately above the channel fader is a SOLO button which, during the time it is engaged, routes the signal of this channel alone to the monitor loudspeaker.

This completes the identification of all controls on the audio input module; replication of this input module with associated faders and switching units completes the control portion of this console.

Overall Control. To the left of the input modules in Fig. 3-16 we see the group of switching pushbuttons previously described. Other controls on the left include talkback, monitor controls for studio and control room, remote controls for four tape recorders, and miscellaneous pushbuttons for the control of other functions.

Jack Field. To the right of the input modules there is a tip/ring/sleeve patch panel, below which are located the 16 output channels plus 4 echo channels and facilities for full

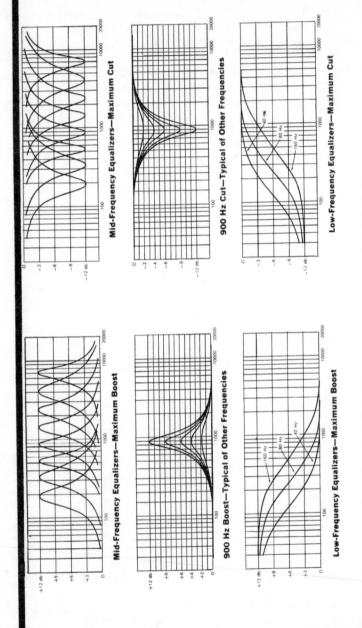

Mid-Frequency Equalizers—Maximum Cut

Mid-Frequency Equalizers—Maximum Boost

900 Hz Cut—Typical of Other Frequencies

900 Hz Boost—Typical of Other Frequencies

Low-Frequency Equalizers—Maximum Cut

Low-Frequency Equalizers—Maximum Boost

H.F. EQUALIZER FREQUENCIES: 10, 12.5 15 kHz   ±2, 4, 6, 8, 12 dB
L.M.F. EQUALIZER FREQUENCIES: 200, 350, 560, 900 Hz ±2, 4, 6, 9, 12 dB

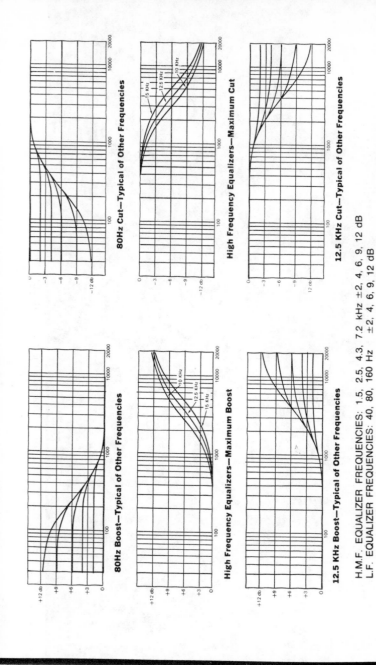

Fig. 3-19 Characteristics of the Audio Designs Series 770 equalizer.

H.M.F. EQUALIZER FREQUENCIES: 1.5, 2.5, 4.3, 7.2 kHz ±2, 4, 6, 9, 12 dB
L.F. EQUALIZER FREQUENCIES: 40, 80, 160 Hz ±2, 4, 6, 9, 12 dB

Fig. 3-20. Audio console manufactured by Rupert Neve, Inc., of Bethel, Connecticut.

simultaneous quadraphonic stereo and monophonic mixdown. This board is equipped with no VU meters, but it does have an oscilloscope-type *Vue-Scan* monitor. Although this level monitoring system is supplied as an integral part of the NRC series boards, it is also available separately.

NEVE CONSOLE

The Rupert Neve Company has established an enviable reputation for quality. Neve consoles with up to 50 input channels and 24 outputs are in use in over 40 countries around the world. A typical Neve console built for recording is shown in Fig. 3-20.

Specifications for Neve professional audio consoles are very close to theoretical limits:

Frequency response: flat within 1 dB, 20 Hz—20 kHz
Output: 26 dBm
Distortion: Not more than 0.075% total at + 20 dBm
Noise: Bus noise not more than −80 dBm
   Microphone equivalent input noise not more than −125 dBm
Crosstalk: Not more than −70 dBm

Specifying distortion at the + 20 dBm level is unusual, but even at such a high level, typical test certificates show less that 0.02% on every channel, which can only be measured on the very best test equipment.[16] Actual crosstalk and noise figures also commonly exceed the specifications.

REFERENCES

1. Lehner, Gerhard. *A Portable Transistorized Mixing Console For the Record Industry*. Jour. Audio Engr. Soc., Vol. 15, No. 1, January 1967 (pp 54—59).

2. Thomsen, Carsten and Alfred C. Antlitz, Jr. *Design of a State of the Art Console for Stereo and Four-Channel Broadcasting*, presented at the 42nd Convention of Audio Engr. Soc., May 1972 (Preprint 860 G—5).

3. Steinke, Gerhard. *A Special Echo-Mixer For a Sound Recording Console*, presented at 16th Annual Meeting of Audio Engr. Soc., October 1964 (Preprint 357).

4. Dilley, William. *New Trends For Studio Console Design*, presented at 17th Annual Meeting of Audio Engr. Soc., October 1965 (Preprint 403).

5. Dilley, William. *A Custom Recording Console*. Jour. Audio Engr. Soc., Vol. 12, No. 4, October 1964 (pp 330—334).

6. Jarvis, John. *A Modular Audio Facilities Mixing System* Jour. Audio Engr. Soc., Vol. 17, No. 1, January 1969 (pp 61—66).

7. Smith, A. Douglas and Paul H. Wittman. *Design Considerations of Low-Noise Audio Input Circuitry For a Professional Microphone Mixer*. Jour. Audio Engr. Soc., Vol. 18, No. 2, April 1970 (pp 140—156).

8. Smith, Allan. *A High-Performance Control Console With Flexibility*. Jour. Audio Engr. Soc., Vol. 18, No. 4, August 1970 (pp 430—435).

9. Blesser, Barry. *An Ultraminiature Console Compression System With Maximum User Flexibility*. Jour. Audio Engr. Soc., Vol. 20, No. 4, May 1972 (pp 297—302).

10. Wiegand, Charles Jr. *An Eight-Foot Board*. **db**. The Sound Engineering Magazine, Vol. 3, No. 3, March 1969 (pp 22—25).

11. Alexandrovich, George. *Multichannel Consoles, Design Trends and Approaches*. **db**, The Sound Engineering Magazine, Vol. 3, No. 6, June 1969 (pp 4, 6, 8).

12. Woram, John. *The Case For Human Engineering*. **db**. The Sound Engineering Magazine, Vol. 4, No. 1, January 1970 (p10).

13. Andrews, R. *New Concepts in Studio Equipment Design*. **db**, The Sound Engineering Magazine, Vol. 4, No. 12, December 1970 (pp 23—25).

14. Alexandrovich, George. *How To Specify A Custom Audio Console*. **db**, The Sound Engineering Magazine, Vol. 4, No. 12, December 1970 (pp 6, 8).

15. Gately, Edward Jr. *Modular Console Design*. **db**. The Sound Engineering Magazine, Vol. 4, No. 4, April 1970 (pp 38—42).

16. Borwick, John. **db** *Visits Rupert Neve*. **dB**, The Sound Engineering Magazine, Vol. 7, No. 6, June 1973 (pp 18, 19).

Ancillary Equipment

4

Judging from the progress of miniaturization, that which is ancillary today may very well be incorporated in the console of tomorrow. It seems only yesterday that equalizers were mounted in relay racks, yet solid-state and active filter techniques have squeezed excellent equalizers with undreamed of flexibility into a few cubic inches and with reduction of cost. It is now not only possible but mandatory to have full equalization available in every input channel.

The console is the control center of the recording activity. But only the controls themselves need be centralized; the device being controlled may be at a distance. The average operator of a multitrack recording studio would likely consider himself incapable of doing his job properly without the numerous pieces of equipment outside of, but controlled by, his console. In this chapter we consider some of these important pieces of auxiliary equipment. In this category are things simply too bulky to be built into a console; but some of the devices represent the growing edge of new electronic developments. Digital techniques, for example, are just appearing in the recording studio for processing sound; we can expect rapid and continuous growth in this area. Integrated circuitry in the form of operational amplifiers is already widely employed in the amplifying and isolating functions of today's consoles, and we can expect rapid growth here as well.

REVERBERATION DEVICES

Reverberation involves tailoring the sound by some method so that the final sound contains the original component plus the same sound delayed a controllable amount to simulate the acoustical conditions of any of a variety of environments.

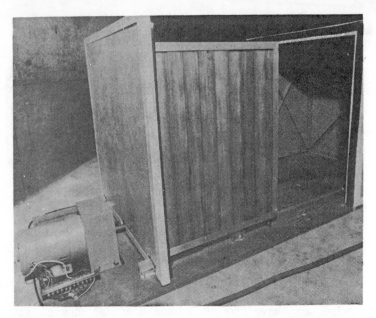

Fig. 4-1. Device for varying the reverberation time of a reverberation chamber. The remotely controlled motor moves the block of glass fiber in and out of the housing, changing the amount of absorption and thus the reverberation time of the room. (Japan Victor Company photo)

Reverberation in its most modest form is used to remove the ultradryness of an acoustically dead studio. At the other extreme, it may be used to simulate the echo effect of an expansive walled-in canyon. Reverberation can be achieved by a variety of means, mechanical or electrical.

Reverberation Chambers

The reverberation *chamber* is surely the brute force approach to obtaining artificial reverberation, yet there are those who insist that the three dimensional reverberation effect of the chamber gives the smoothest and most natural reverberation effect available and are willing to pay for it. Detractors point out the difficulty of remotely adjusting the reverberation time in such chambers. The Japan Victor Company came up with a neat solution to this problem.[1] The firm has six chambers in its Tokyo studios (in addition to 14 plates and six magnetic disc reverberators), each about 3000 cu ft in volume. Each chamber contains three remotely controlled acoustic absorbers (Fig. 4-1). A 7 cu ft glass fiber block is withdrawn from or pulled into a plasterboard box by a motor controlled from the consoles. This variable absorption gives a broad range of reverberation times. The solid mass of glass fiber can easily be built up from thicknesses of fairly rigid glass fiber board such as Owens Corning Type 700 industrial insulation.

Other chamber designs are available,[2] but space considerations and the advent of practical miniature electronic circuits seem to be supplanting their use in modern recording studios.

Reverb Plates

Next to the reverberation chamber in the production of lifelike reverberation is Dr. Kuhl's reverberation plate. The latest model is the EMT 140 TS of Fig. 4-2. The operation of this device depends on the bending vibrations set up by a transducer in a hefty steel plate supported by a tubular steel frame in a box approximately 8 × 4 × 1 feet, weighing about 400 lb. The vibrational energy set up in the plate is reflected from the edges, much as sound is reflected from the walls of a concert hall, except that it is basically a two- rather than a three-dimensional effect. It is difficult for the human ear to detect the difference between these two conditions, however.

Reverberation time is variable from 1 to 4 seconds by a second plate of highly porous material mounted parallel to the steel plate. A hand wheel or remotely controlled motor swings this damping plate toward or away from the steel plate to vary the effective reverberation time (the plates never touch). The variation of the reverberation time with frequency depends upon the setting of the damping plate, according to Fig. 4-3.

Fig. 4-2. The EMT 140 TS reverberation plate suitable for both mono and stereo. (Gotham Audio Corporation photo)

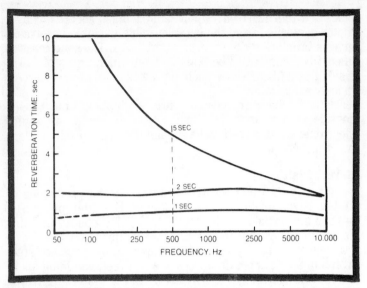

Fig. 4-3. Reverberation time vs frequency characteristics of the EMT 140 TS reverberation plate. (Gotham Audio Corporation)

The device is usable for both mono and stereo. For stereo, Fig. 4-4, part of the unreverberated signals of channels A and B are combined to form a composite mono signal containing all the informational content of the stereo channels, and this is applied to the plate. The two contact microphones, placed at unequal distances from the driving coil, pick up statistically distributed signals which are fed to their respective channels.

Reverb Foil

The EMT plate, which has been the elegant standard of the recording industry for many years, seems to be in the process of being replaced by the EMT 240 reverb foil of Fig. 4-5, which is one-third the weight and one-fifth the volume of the standard plate. The size and other characteristics make the foil much more adaptable to mobile recording. The special gold alloy foil is only about 10.6 by 11.4 inches with a thickness of 18 microns (18 millionths of a meter). Both components of the alloy are deposited simultaneously by an electrolytic technique. Again, the variation of the reverberation time is achieved by changing the proximity of a porous absorbent material. Figure 4-6 shows the greater uniformity of reverberation with frequency as compared to the plate, at least for the midrange frequencies to which the ear is most sensitive. The most

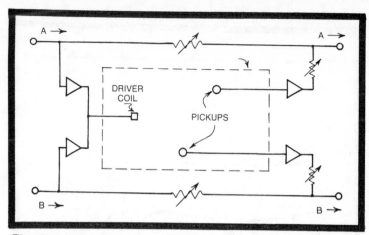

Fig. 4-4. Method of using the EMT 140 TS plate for stereo reverberation. (Gotham Audio Corporation)

Fig. 4-5. The EMT 240 reverb foil utilizing a very thin foil of special gold alloy as the reverberation element. (Gotham Audio Corporation photo)

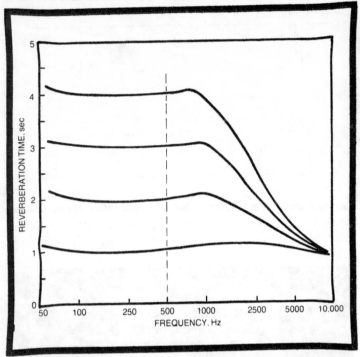

Fig. 4-6. Reverberation time vs frequency characteristics of the EMT 240 reverb foil. (Gotham Audio Corporation photo)

important aspect of such a characteristic is that it be free of peaks which impart colorations to the reverberated signal.

Reverb Spring

Artificial reverberation units utilizing the one-dimensional principle are available in profusion, many of questionable quality in very inexpensive form for amateur use. Most such units utilize a spring to get a considerable length of wire in limited space. A professional reverberator based on this principle is the AKG Model BX 20E unit; it is shown without its protective housing in Fig. 4-7. This spring is excited in a torsional mode. Of course, this delay line isn't just any piece of bed spring. It is very carefully engineered to provide the proper characteristics by local etching of the wire, deformation of certain turns of the spring, variation of the spring diameter with length, and placement of damping elements along its length. The net result is uniform reverberation throughout the audible frequency range, as shown in Fig. 4-8.

88

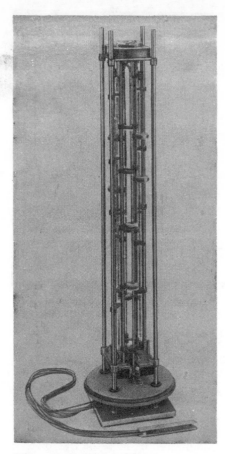

Fig. 4-7. The AKG BX 20E reverberation unit utilizing a spring excited in torsional mode.

Magnetic Reverberators

Some of the earliest reverberators made use of the magnetic recording principle, either on disc or tape. The signal to be reverberated is recorded on the magnetic medium; several playback heads, distributed downstream, pick up the signal with varying degrees of time delay from the original. Combining these signals gives something that might pass for reverberation in noncritical applications, but it is really not reverberation in the strict sense, only a series of delayed sounds more like echos. If many pickup heads are spaced very closely and the relative levels are carefully adjusted, a reverberation-like quality is approached.

Because of its fragility and the service problems going along with any mechanical system, magnetic reverberators are not often seen in modern professional circles, though they enjoyed a spurt of popularity in the late 1950s.

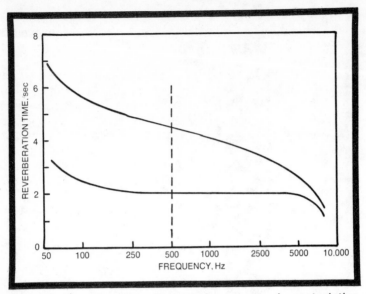

Fig. 4-8. Reverberation time vs frequency characteristics of the AKG BX 20E spring unit.

NOISE REDUCTION SYSTEMS

Noise can be a serious problem in any audio system, but it is especially so in multitrack recording. Every track added increases the final mixed down noise level. This is over and above the fact that reduction in track width to accommodate so many tracks in a given width of tape also reduces the signal-to-noise ratio.

Usable dynamic range is limited by noise on the low end and by distortion at high levels. Even if the electronic circuits were perfectly noise-free, which they most decidedly are not, our magnetic tape medium presents its own private limitations in inherent noise (tape hiss) at low levels and tape saturation characteristics at high levels. We are obligated to operate within these boundaries imposed upon us, but we are not entirely helpless: There are things we can do to survive in the recording business with critical demands of the public being what they are.

The effective noise level of a recording system may be reduced by many methods, ranging from simple to sophisticated, from the relatively ineffective to the dramatically effective. A simple high-pass filter can reduce rumble, a low-pass filter can reduce hiss and scratching, and even the lowly tone control used as a low-pass filter can give some modest reduction in apparent noise. The phrase, "the narrower the window, the less muck comes through," describes the effect of bandwidth. But there is a price to pay in loss of program content in the use of such gross approaches to noise reduction.

Recording engineers have always been confronted with the problem of squeezing the great dynamic range of an orchestra, for example, into the limited dynamic range of their system by the simple expedient of manually increasing gain on low passages and reducing gain on high passages. If equal skill could be assured in the reproduction of this sound so that complementary gain changes reestablish the original dynamic range, all would be well. This can be done electronically with compressors and expanders. These are built with fast attack time (a few milliseconds) and much slower recovery time (a few seconds) to avoid audible modulation of program content. The main problem with such a system is in the breathing or pumping effect as the background noise level changes. This adverse effect is not always noticeable but sometimes, such as during a low-level instrumental solo with occasional silent portions, it is very objectionable; and for this reason (among others) it has never been widely applied to recording.

The Dolby System

The noise reduction system devised by Dr. Ray Dolby and first marketed about 1967 is basically a compression/expansion scheme.[3,4,5,6] Dolby added a vital ingredient to the long line of noise reduction systems which

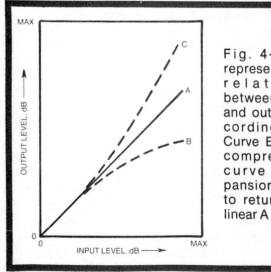

Fig. 4-9. Line A represents a linear relationship between the input and output of a recording channel. Curve B represents compression and curve C the expansion necessary to return B to the linear A condition.

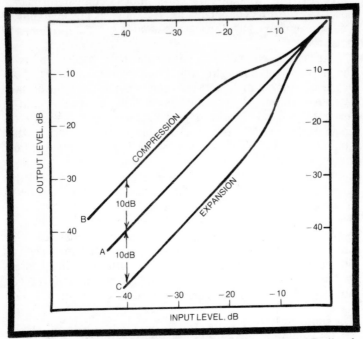

Fig. 4-10. The input—output characteristics of the Dolby A noise reduction system. The compression and expansion are limited to low-level signals.

preceded him—he divided the recorded spectrum into four separate bands and operated on each independently. During recording four compressors are used; during playback, four expanders provide complementary action. The "breathing" problem is eliminated by his splitting of the four bands and by use of precisely paced compressor and expander action. Most importantly, the Dolby system operates only on soft passages rather than continuously.

In Fig. 4-9, line A represents the linear relationship between input and output of a recording channel for which we continually strive. In the old system, curve B represents the compression and curve C the subsequent expansion required to bring curve B back to the linear relationship A.

To show the Dolby action at low levels, it is more convenient to graph the relationship between input and output as in Fig. 4-10. The linear line A is our ultimate goal. Compression according to curve B in recording is compensated for by the complementary expansion of curve C in reproduction, bringing the resultant to the linear condition

Fig. 4-11. A typical multitrack Dolby noise reduction installation of 24 Model 361 units.

A. Typical multitrack Dolby installations are shown in Figs. 4-11 and 4-12.

Table 4-1 summarizes the pertinent data of Dolby action. The four bands are a compromise between cost and

Fig. 4-12. The Dolby M-16 noise reduction system providing 16 channels with built-in record—play changeover facilities.

effectiveness and each has its effect on specific noise components. Band 1 operates primarily on hum and rumble; band 2 on broadband noise, crosstalk, and printthrough; bands 3 and 4 are primarily effective in reducing tape hiss and tape modulation noises. The compression/expansion ratio is 10 dB, except for band 4, which has 15 dB to better fit the tape noise spectrum. With average orchestral music, band 2 is compressed most of the time while the other bands are operative less frequently.

The genius of the Dolby system is most evident in the decision to operate on low-level signals. After all, noise interferes most with low-level music signals. Any Dolby tracking errors that might exist are at such a low level that

Table 4-1. Dolby Noise Reduction System

Band	Frequency Range	Repressed Noise	Compression	
			Ratio	Frequency
1	80 Hz lo pass	Hum, rumble	10 dB	Fairly often
2	80 Hz-3 kHz	Broadband noise, crosstalk, printthrough	10 dB	Almost all the time
3	3 kHz hi pass	Tape hiss, modulation noise	10 dB	Fairly often
4	9 kHz hi pass	Tape hiss, modulation noise	15 dB	Rarely

they are unobtrusive. This makes the system tolerant of gain errors, a characteristic most valuable in stereo because no control interconnections are required. A very important byproduct of low-level compression and expansion is the fact that the Dolby units look like unity-gain amplifiers at the high levels at which calibration and equipment lineup procedures are carried out.

The use of 8-, 16-, or even 24-track tape machines with high standards is made practical only by the use of some system for reducing noise. The *Dolby A* system (as contrasted to the simplified *Dolby B* used in consumer equipment) has become the virtual standard for at least the more affluent of the recording industry around the world. This system gives more than 10 dB improvement in signal-to-noise ratio across the entire band—in other words, it reduces hum and other low-frequency noises as well as high-frequency hiss.

The DBX System

A later arrival on the noise reduction scene is the DBX system[7] (the firm uses lowercase letters for its logo), totally incompatible with the Dolby system. A 16-channel DBX unit is shown in Fig. 4-13. Like the Dolby, the DBX is a code/decode system depending on compression and expansion for its operation, but there the similarity ends. Dolby divides the audible spectrum into four bands, operating on each in a different way. DBX operates on the entire spectrum as a unit. Dolby confines compression and expansion to low-level signals, DBX applies a uniform factor.

The DBX success in reducing noise is achieved by a series of signal processing operations starting with high-frequency preemphasis. This reduces the effect of tape hiss varying wth the signal envelope. To get adequate transient response and tracking properties, advantage is taken of recent analog computer techniques to obtain a true RMS level sensor. The result: some 30 dB of broadband noise reduction is attained in equipment that is easy to handle in the control room as level matching and pilot tone have been eliminated. Each DBX unit contains a code and a decode circuit making remote control unnecessary.

There are other noise reduction systems, such as the Burwen 2000,[8,9,10] and Robert Orban has recently proposed another program-controlled noise filter.[11] Space will not be given to them here, but the interested reader can refer to the literature. Digital processing techniques promise to render the present systems obsolete in the future.

Fig. 4-13. The DBX Model 216 multichannel record and play noise reduction system.

COMPRESSORS AND LIMITERS

In the preceding section we have considered compressors and expanders used by Dolby and others to achieve noise reduction. Let us now consider the compressor and limiter as devices which have their own specific contribution to sound recording.[12-17] The compressor may be defined as an amplifier whose gain is decreased as the program level increases, averaged over a time interval. This interval is generally long compared to the natural fluctuations of program content. Thus, the compressor affects the average program level. Very short program peaks can "get through" the control circuitry of the compressor before it has time to operate, resulting in overdrive of systems which follow. The limiter, designed to control these peaks, has a very fast-acting circuit which controls the peak excursions missed by the compressor.

We see, then, that both compressors and limiters are circuits controlled by program level—the compressor increasing the average program level, and the limiter eliminating system overdrive that might otherwise be caused by short-duration peaks. These two functions, often combined in the same equipment, may be applied to individual channels or to the mixed sound, or both. Solid-state techniques have made possible reduction of size so that by inclusion in input modules of the console, or by remote cards controlled from the console, compression and limiting of the signal of each channel are becoming available.

The action of the compressor is illustrated in Fig. 4-14. The straight line with compression ratio of 1:1 represents a linear relationship (in decibels) between input and output. A compression ratio of 2:1 means that for each 2 dB of input change, only 1 dB change of output level results. The various compression ratios are switch selectable and hinge about the intersection point of normal input reference level and normal output reference level. A compression gain control allows desired compression to be selected.

Limiter action is illustrated in Fig. 4-15. The limiter circuit has no effect until the input signal reaches the limiting threshold. This threshold is normally adjusted to be at a very high level, just short of the upper limit of the system, whether it be the tape's saturation level or amplifier clipping point. Any program peaks exceeding this threshold value (input higher than −35 dB or output higher than +15 dB in the example of Fig. 4-15) have their top excursions limited. Limiter setting A would be no limiting at all, a linear relationship between input

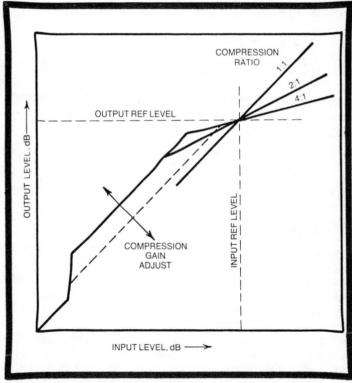

Fig. 4-14. Compressor action graphically illustrated.

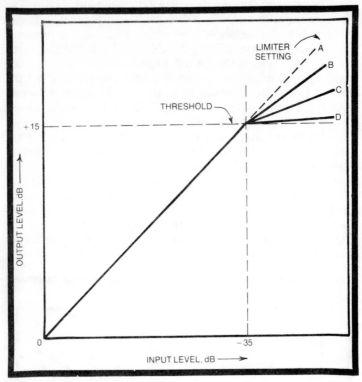

Fig. 4-15. Limiter action graphically illustrated. Limiting action increases from A to D.

and output. Setting B would give only a slight limiting action; setting C considerably more, and setting D would give maximum limiting, allowing only slight excursion beyond the +15 dB output level. Manufacturers of limiters find the word *clipping* repugnant because square corners resulting from chopping off peaks mean that distortion products are generated. However, any degree of limiting changes the waveform to some extent and thus introduces distortion. In the tradeoff between the controlled distortion introduced by the limiter and that which would be introduced by overdriving the system, the limiter comes off as a very valuable and useful tool in multitrack recording.

Compressors and limiters differ greatly in the dynamics of attack and release. The compressor attack time is generally in the 1-10 millisecond region, but the release time is purposely much longer. If compressor release time is ridiculously long, a program peak reduces the gain and it stays reduced. At the other extreme, if the compressor release time is very short, each element of the program is continuously subjected to gain changes, resulting in a very unpleasant effect. Commercial compressors have an adjustable release time ranging from a few tenths of a second to several seconds. Limiter attack time is made as fast as possible, of the order of 1 to 5 μsec, to prevent a rapidly rising peak overdriving the equipment which follows. Limiter release time is usually several seconds to minimize harmonic distortion associated with fast release times.

EXPANSION AND GATING

The dynamic range of a signal following an expander is greater than the signal fed into its input. An expansion ratio of 2:1 means that for a 1 dB change of level at the input, a 2 dB change of level appears at the output of the expander. Figure 4-16 illustrates how expansion, limiting, and gating functions can be combined in one instrument. The expansion feature can restore some of the dynamic range lost in a compressor earlier in the audio chain. It is important that expansion does not begin below the noise level of the system.

The *Kepex 500* (**k**eyable **p**rogram **ex**pander), manufactured by Allison Research and pictured in Fig. 4-17A and B, can be used as a gating device. A signal falling below a predetermined level is, in effect, turned off. This can be a very useful feature around a multitrack recording studio. For example, let us assume that less than perfect circumstances during recording resulted in magnetic tracks that do not have very good separation, and that spreading and spillover from one track to the other is too great. This instrument can be used to reduce or eliminate interfering sounds such as air-conditioner rumble, aircraft, and traffic noise; it can even dry up a signal recorded in a highly reverberant room and take care of the spreading at the same time. There are many occasions when using the gate feature might save the day. The sensitivity levels can be set so that interfering lower level sounds are cut off, leaving only the desired higher level sounds of the instrument or vocalist. Such devices can be very valuable during both recording and mixdown. Not all channels or tracks are busy all the time. Gating will automatically shut off any channels or tracks on which the signal level falls below a certain level, eliminating extraneous low-level noise from microphones and hiss from tape.

A measure of control of even room coloration and reverberation can be effected by gating. Sounds from air conditio1ers, sirens, chair scraping, turning pages of the

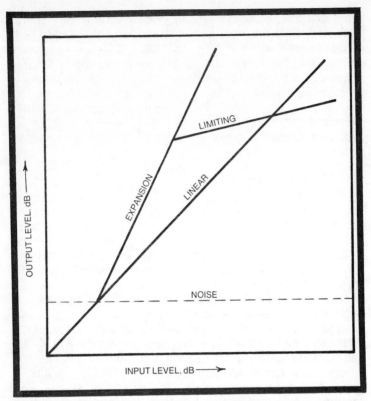

Fig. 4.16. Expansion, limiting, and gating functions may be combined in one instrument.

score, etc., can also be reduced or eliminated. While such gating devices can turn channels on and off automatically, and thus relieve the operator of this chore so that he may concentrate on the bigger job, it must be emphasized that they add that many more adjustments that must be made very carefully or serious distortion can result.

PROGRAM LEVEL INDICATORS

The venerable standard for keeping tabs on the program level for many decades has been the VU meter. The term VU is an abbreviation for *volume units*. For practical purposes, a VU is the same as a decibel. The 0 VU point is equivalent to 0 dBm; that is, 0 VU is 1 mW of power in a circuit of 600Ω impedance. The important characteristic of the standard VU meter, apart from indicating level, is its dynamic response as the reading on highly fluctuating program material is very much affected

A

B

Fig. 4-17. The Kepex 500 can be used as a gating device to reduce or eliminate interfering sounds from a signal. (Allison Research, Inc. photo)

by the attack and release times built into the instrument. Let us simply say that not every meter in consumer equipment with VU printed on its face has the important dynamic response of the standard instrument.

Modern consoles have traditionally used VU meters as a means of indicating levels of the various channels. An array of 8 VU meters begins to pose a monitoring problem for the

Fig. 4-18. The Vue-Scan program level indicator can present 20 or 28 channels of audio level information in small space on a TV-type screen. (Audio Design & Manufacturing photo)

operator, but as the number of channels increases to 16 or 24, the problem gets to be quite serious. Obviously, new forms of program level indicators are demanded. Accordingly, we are now seeing the forerunners of advanced devices appearing in some of the later consoles and offered as an ancillary unit for existing consoles equipped with VU indicators.

The *Vue-Scan* program level indicating device offered by Audio Designs (Fig. 4-18) makes relatively easy the visual monitoring of 16 channels of audio information. Based on digital and analog techniques, the Vue-Scan device presents the level information of each channel as a bar graph on a TV-type monitor screen. The lower two-thirds of the monitor screen area is blue, and the upper one third is red. As peak levels go into the red area, the increased height and intensity are visually quite demanding. The dynamics may be adjusted from the front panel either for peak indicating or to match VU ballistics. The range of each bar indicator is from −20 to +8 dBm and, with an extender, can be increased. Levels up to +4 dBm (or +10 dBm with extended range) appear in blue while all levels above appear in red. Vue-Scan monitors are available to monitor as many as 28 channels.

Another promising replacement for the VU meter is based on the light-emitting diode (LED). In the simplest arrangement, a single LED is made to light when the program signal level exceeds a preset threshold. In a more complicated arrangement a stack of LED devices may be adjusted to have their firing thresholds a fixed "distance" apart (2 dB, for example), so that an indication of absolute level can be obtained. Such circuits need to incorporate an LED light-pulse lengthening feature so that extremely short transients can be readily seen. Another arrangement is to imbed a series of LED indicators on an arc in the face of the VU meter so that information on both average and peak signal level are available simultaneously.

FOLDBACK

It is necessary to provide foldback (cue) signals from the console to artists in the studio. This is particularly important during overdub sessions and, if acoustic separation of performers is great, it may also be required during normal recording sessions to enable certain musicians to hear others. The usual method is to string headphone lines across the studio to the artists, as required, and plug in headphones as needed. This is a most elementary approach, but it is effective. It suffers from one major drawback—just that many more cables to add to the maze of microphone and electric musical instrument cables already making the studio floor look like a snake farm.

There is an ancient method which can be used—better known in Europe than in the United States: the induction loop. With this method, the artists are completely free to move about, receiving their foldback without entangling wires. The induction loop principle is illustrated in Fig. 4-19. The foldback signal from the cue bus is fed into an amplifier capable of providing 3 to 5 watts. It is possible that the foldback amplifier is directly capable of doing this. The output of this amplifier is connected to a loop of several turns of wire around the room through an impedance-matching transformer. The loop may be placed anywhere around the periphery of the studio—at floor level, around the wall, or hidden above the ceiling. It could also encircle the control room if desired for adjustment purposes. Current flowing in this large loop produces a magnetic field, the magnitude of which depends upon the number of turns and the amount of current flowing in them. There is reasonable uniformity of flux throughout the studio but the flux drops to zero near the loop. The flux increases

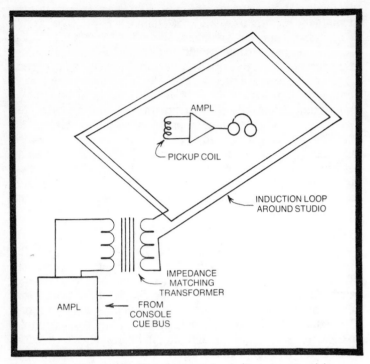

AMPL

PICKUP COIL

INDUCTION LOOP
AROUND STUDIO

IMPEDANCE
MATCHING
TRANSFORMER

AMPL ← FROM
CONSOLE
CUE BUS

Fig. 4-19. The induction-loop principle can provide cue signals to a studio without trailing headphone cables.

outside the loop, which might interfere with the use of another loop in an adjoining studio.[18] The Murdock system is said to make adjacent-room loop operation practical.

The signal is picked up by a small coil which feeds a tiny battery-operated amplifier affixed to and driving the headphones. The person wearing the headphones can hear the signal fed into the loop with no constraining wires to drag around. A headphone equipped with such an amplifier and pickup coil is illustrated in Fig. 4-20. This unit, manufactured by Beyer and distributed in the U.S. by Revox, has a battery life of 150 hours; total weight of headphones and amplifier is only about 8 oz. A number of American firms also offer such units for classroom use, chiefly for teaching children with impaired hearing.[19] Although applications of the induction loop for recording studio foldback are not too common, it is suggested here as a possible solution in certain situations where artist mobility is a prime consideration. Electrical noise from such things as fluorescent lamps within the loop may be a problem, but if fluorescent ballast reactors are the

104

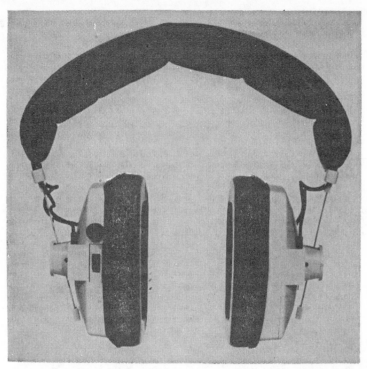

Fig. 4-20. The Beyer DT 100V headphones can pick up the signal from an induction loop without headphone umbilicals. (Revox Corporation photo)

source, they should be banished to a distant closet anyhow because of their intolerable and unpardonable noise output.

REFERENCES

1. Shiraishi, Y., K. Okumura, and M. Fujimoto. *Innovations in Studio Design and Recording in the Victor Studios*, Jour. Audio Engr. Soc., Vol. 19, No. 5, May 1971 (pp 405–409).

2. Rettinger, Michael. *Reverberation Chambers For Recording*, Jour. SMPE, Vol. 45, No. 5, November 1945 (p350).

3. Dolby, Ray. *An Audio Noise Reduction System*, Jour. Audio Engr. Soc., Vol. 15, No. 4, October 1967 (pp 383–388).

4. Eargle, John. *The Dolby Noise Reduction System—Its Impact on Recording*, Electronics World, May 1969 (pp 32–34).

5. Borwick, John. *Dolby Revisited*, db, The Sound Engineering Magazine, Vol. 3, No. 8, August 1969 (pp 20–22).

6. Woram, John. *Interfacing a Dolby System*, db, The Sound Engineering Magazine, Vol. 6, No. 10, November 1972 (pp 22–25).

7. Blackmer, David. *A Wide Dynamic Range Noise-Reduction System*, db, The Sound Engineering Magazine, Vol. 6, No. 8, August/ September 1972 (pp 54–56).

8. Burwen. Richard. *Design of a Noise Eliminator System*. Jour. Audio Engr. Soc.. Vol. 19. No. 11. December 1971 (pp 906—911).

9. Ford. Hugh. *Review: Burwen 2000 Noise Eliminator*. Studio Sound. Vol. 16. Nos. 2/3. February/March 1974 (pp 49. 50 and 52. 53).

10. McKenzie. Angus. *Field Trial: Burwen 2000 Noise Eliminator*. Studio Sound. Vol. 16. Nos. 2/3. February/March 1974 (pp 59/60). [See also *Studio Sound*. Vol. 16. No. 5. May 1974 (pp 31—33) for Burwen's rebuttal.]

11. Orban. Robert. *A Program-Controlled Noise Filter*. Jour. Audio Engr. Soc.. Vol. 22. No. 1. January/February 1974 (pp 2—9).

12. Blesser. Barry. *An Ultraminiature Console Compression System with Maximum User Flexibility*. Jour. Audio Engr. Soc.. Vol. 20. No. 4. May 1972 (pp 297—302).

13. Noble. James and Robert Bird. *A Dual-Band Audio Limiter*. Jour. Audio Engr. Soc.. Vol. 17. No. 6. December 1969 (pp 678—684).

14. Woram. John. *The Sync Track*. **db**. The Sound Engineering Magazine. Vol. 4. Nos. 10 and 11. October/November (pp 25—28 and 16. 18).

15. Buff. Paul. *A Combination Limiter*. **db**. The Sound Engineering Magazine. Vol. 6. No. 2. February 1972 (pp 22—24).

16. Conner. Dale and Richard Putnam. *Automatic Audio Level Control*. Jour. Engr. Soc.. Vol. 16. No. 3. July 1968 (pp 314—320).

17. *Forum Compressor and Limiter Guide*. **db**. The Sound Engineering Magazine. Vol. 4. No. 10. October 1970 (pp 31—35).

18. Bosman. D. *Magnetic Induction Loops With Geometrically Restricted Fields For Communication in Schools for the Hard-of-Hearing*. presented at the 44th Audio Engr. Soc. convention. Rotterdam. February 1973 (Preprint D-2R).

19. Beyer products are handled by Revox Corporation. 155 Michael Drivel Syosset. New York 11791. Murdock CH-4 cordless headphone systems are handled by Instructional Materials and Equipment Distributors. 1520 Cotner St.. Los Angeles. California 90025. Information on the Scintrex Model L-100 induction loop headset may be obtained from Scintrex. Inc.. 400 Creekside Drive. Amherst Industrial Park. Tonawanda. New York 14150.

Multitrack
Recorders

5

Multichannel consoles and multitrack recorders may be considered the twin handmaidens of separation recording. The multitrack recorder gives us a permanent storage of the separate elements of the musical group upon which subsequent mixdown flexibility depends. As far as fundamental operating principles are concerned, if you've seen one single-track magnetic recorder, you have seen them all. However, the path from full-track mono on ¼-inch tape to 40 tracks on a 2-inch tape is littered with frustrated hopes and shattered dreams.

All magnetic recorders/reproducers have some form of magnetic material moving past heads which perform erase, record, and playback functions. In the tape recorder of special interest to us it is convenient to refer to two systems, the tape transport and the electronics. This distinction is not as clear today as it was in the earlier days: Electronics is almost as much a part of today's tape transport mechanisms as it is of signal amplification, generation of bias and erase currents, etc. Logic and switching functions alike are more and more becoming solid-state replacements for their mechanical counterparts. Nonetheless, there is still a logical distinction between the mechanical and the electrical aspects of a tape machine.

The specialized multitrack recorders are only more sophisticated extensions of the simple principles illustrated in Fig. 5-1. Tape driven by a capstan moves from a supply reel to a takeup reel. On the way the tape passes over an erase head, a record head, and a playback head. An oscillator operating at a frequency at least three times as high as the highest audio frequency being recorded erases any material previously

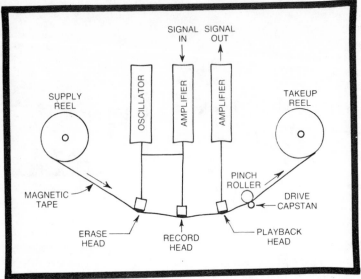

Fig. 5-1. Basic magnetic recorder/reproducer configuration.

recorded on the tape. The signal to be recorded is amplified and fed to the record head mixed with some of the high-frequency alternating current from the oscillator. This is called *bias*, and in conventional recorders one oscillator usually supplies both erase and bias current, although they are often separated in the better machines to optimize performance.

Without going into all the details of the magnetic recording process, an audio signal is converted to magnetic flux variations of the tape's magnetic coating as the tape passes the record head. As the tape passes the playback head, the magnetic flux variations induce an alternating current in its winding which, after amplification, is the output signal. Note that it takes time for the tape to move from the record head to the playback head; the signal out is thus delayed from the signal going in. This is the familiar delay encountered when comparing input and output signals during the recording process.

One of the aspects of recording which is hidden and often taken for granted is that equalization is needed to compensate for the characteristics of the magnetic recording process. Such equalization is built into a machine, so the operator rarely needs to concern himself with its details. Considering Fig. 5-1 as a full-track ¼-inch magnetic recorder/reproducer, we have established a foundation for our discussion of more complex machines.

In going from 2 or 4 tracks, common in consumer and professional recorders, to 8, 16, or 24 tracks is a small step in basic principles but a large step in complexity and sophistication. The uses to which the multitrack recorder is put brings demands for high mechanical precision, highly stable and predictable electrical performance, and flexible operation. A good example of this is the drive mechanism. The synchronous motor driven from the power mains directly is rapidly giving way to the servo-controlled drive, which gives not only extremely constant speed with low wow and flutter, but the ability to change the speed by a simple change of electronic components rather than mechanical means. A variable-speed feature is valuable for special effects, correction of timing errors, and salvaging of off-key musical tracks. Most important, however, is the ease of synchronizing one servo-controlled multitrack recorder with other tape machines or with film and video equipment. Although not all multitrack machines currently have servo control, it is definitely a feature of the future for the better quality recorders.

SIGNAL-TO-NOISE RATIO

In the ¼-inch full-track magnetic recording, still widely used in monophonic broadcasting, the pickup head gap "reads" almost the entire width of the tape, about 234 mils of the 250 mils width. This is about as wide as they get, for the greater the number of tracks, the narrower they must be. Going to tapes as wide as 2 inches to meet the demand for more and more tracks has never resulted in tracks as wide as the old ¼-inch track.

But there no longer exists the need for wide recording-path swaths. Tapes have been improving steadily since their introduction. And with each improvement in coating uniformity comes a complementary decrease in track width to achieve the same total sound. Head technology has similarly been keeping pace, with the result that a 25-mil track width today can do the same job as a ¼-inch track of the 1950s—better.

In the multitrack recorder, we are interested in the width of the individual tracks because, for a given tape speed, the signal magnitude is proportional to it. If the tape speed is cut in half the signal-to-noise ratio is reduced 3 dB. Cutting track width in half also reduces signal-to-noise ratio by 3 dB.

Specifying the signal-to-noise ratio of a multitrack magnetic recorder/reproducer for ready comparison with competitive machines is complicated by the great number of variables involved. If weighted, what standards? What is the signal reference level: +3 dBm, 3% total harmonic content? What tape is used? What tape speed? Width? Number of tracks? To get an idea of the signal-to-noise ratios realized commercially it is necessary, because of the variables, to compare products from one company. Table 5-1 is a tabulation of the noise specifications of the several models manufactured by Stephens Electronics, Inc.

In Table 5-1 all noise levels are expressed in reference to the number of decibels below a maximum signal determined by the point at which 3% total harmonic distortion occurs. For the 16-track 15 ips (inches per second) case, the noise is 66 dB below this reference level. From Table 5-1 we note a 3 dB lowering of the noise by increasing the speed from 15 to 30 ips. We also note a 4 dB increase in noise by going from a 16-track channel to a narrower 40-track channel. That tape hiss is the important factor is shown by the significantly lower noise when the tape is stopped.

CROSSTALK

The closer the spacing between heads in a given head stack, or the closer the head stacks, the greater the problems of signals in one track feeding through to adjacent tracks. In recording head stacks there is magnetic coupling and transformer action between the individual heads, principally adjacent ones. This may be helped by designing for maximum physical separation (hard to get in multitrack heads) and by the use of high-permeability magnetic shielding. In reproducing-head stacks there is leakage primarily between windings of adjacent heads and a fringing effect which is especially troublesome at the lower audio frequencies. These effects result in so-called crosstalk.

The seriousness of interference between two adjacent tracks depends, among other things, upon the compatibility of signals on the two tracks. If the signals have no relationship to each other, the interference effect is greatest. In the use of the multiple tracks for recording music as we have discussed in

Table 5-1. Signal-to-Noise Specifications

No. tracks	at 15 ips	at 30 ips	with tape stopped
16	66	69	74
24	63	66	71
40	62	65	70

this book, the signals on the different tracks are related. In fact, the acoustical separation between instruments in the studio is far from perfect. Realizing that interchannel separation in the studio of 20 dB is about as good as we can expect, we are prepared for the bad news regarding intertrack separation on the tape. This is a subject well camouflaged in most recorder specifications, and figures for comparing competitive machines are difficult to find.

Crosstalk is usually measured by recording on one track at a certain level and measuring the signal picked up by the reproducing head of an adjacent track, then expressing the level difference in decibels. The specification sheet for the Studer A80 16-track machine (manufactured in Europe and distributed in the U.S. by Willi Studer of America) lists crosstalk rejection of 40 dB or more, 60 Hz—15 kHz. Figures of 50 to 60 dB show up for American machines, if listed, but usually are not listed; and when they are given, there is a tendency to do so for a single frequency in the most favorable part of the spectrum, avoiding the low and high audio frequencies where crosstalk is worst.

The Studer specifications include one other bit of data bearing on separation problems of multitrack recorders. In another section the sync procedure will be discussed in detail; at the moment, however, all we need to know is that we are here considering crosstalk between heads *in the same stack.* The Studer specifications, complete as they are, also give us some information on the amount of separation between adjacent tracks in the same stack. "Crosstalk rejection between record channel and any adjacent sync channel at 15 ips: 26 dB or more at 1 kHz, 10 dB or more at 10 kHz." So we are working with recording/reproducing equipment having intertrack separation of the same order as the acoustic separation we are able to manage in the studio.

CAPSTAN DRIVE: FIXED AND VARIABLE

The tape in a magnetic recorder/reproducer is driven by a capstan rotating against a pliable or other pinch roller with the tape between the two. The quality of the recording depends upon the uniformity and constancy of tape motion. If the tape speed fluctuates slightly, the signal recorded on the tape will be frequency modulated and tones that should be steady when played back will fluctuate in frequency as the speed fluctuates. Herculean engineering effort has gone into improvement of tape drive systems and, in large measure, the quality of a recorder can rise no higher than the quality of its drive system.

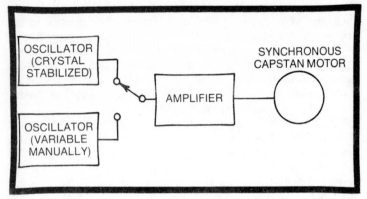

Fig. 5-2. With this arrangement, both an extremely constant and a variable capstan speed may be obtained.

Synchronous motors, whose speed is determined by the very constant frequency of the power source, have been the dominant professional recorder drive for many decades. The hysteresis synchronous motor has been widely used, especially where interlock of several machines was necessary. The problem with synchronous motors is lack of flexibility. Although synchronous motors with two sets of windings have been used to provide two tape speeds, the speed of the synchronous motor is fixed, because the frequency of the power source is fixed.

One way of achieving needed flexibility with a synchronous motor is illustrated in Fig. 5-2. Instead of obtaining alternating current from the power mains, a local stabilized oscillator supplies the alternating voltage of an extremely constant frequency which is amplified to provide sufficient power (perhaps 100 watts) to drive the motor. The synchronous motor then faithfully holds its speed as constant as the frequency of the oscillator, which may be crystal controlled. A variable-speed feature is available by switching to another oscillator whose frequency can be manually controlled by plus or minus 25% or so. This makes possible the correction of tracks recorded at an unstandard speed as well as certain special effects.

Servo techniques are being widely applied to recorder drives because of the precision and flexibility which characterize them. A servomechanism has been defined as a power amplifying device in which the amplifying element driving the output is actuated by the difference between the input and the output. Perhaps that opaque statement can be clarified with an example.

In Fig. 5-3 the DC capstan motor is driven by the servo amplifier. Associated with the motor is some sort of speed sensor which feeds information on the actual motor speed back to the comparator. As its name implies, the comparator compares this sensor signal with some standard (possibly a crystal-controlled oscillator) and sends an error signal back to the driver, which controls the power to the motor in a direction to bring the motor speed to agree with the standard. Through the speed sensor signal and the feedback circuit the motor speed is automatically held to the standard value with great accuracy. If it is desired to increase or decrease the motor speed, it is only necessary to override the standard signal with a manually controlled signal. All of the sensing, comparing, and error aspects of the feedback loop are accomplished at low power levels, the servo amplifier being the only relatively high-power circuit.

A very important feature of the servo system is the ease with which tape transports can be locked together; thus, two 8-track recorders can easily be synchronized to provide a 16-track recording or playback capability, or two 16-track machines to provide 32 tracks with absolute synchronization between them. Similarly, locking in with compatible videotape or film equipment becomes a routine operational matter.

Servo systems can be applied to supply and takeup reel drive systems as well. In this case, tension of the tape is the crucial factor, and the sensing device must feed tension information back to the comparator. Electronic braking also becomes simple and practical through the servo system. With

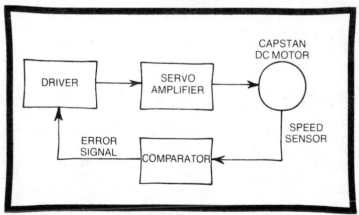

Fig. 5-3. Servo speed control for capstan drive is achieved by comparing the actual speed of the motor against some standard and feeding back a correcting error signal.

a DC servo motor, the torque is smooth and steady, which tends to lower the flutter of the servo-driven machine.

CONTROL LOGIC

Relays are still commonly used in switching tape recorder functions but integrated logic circuits are fast appearing. Solid-state switches are destined to replace electromechanical switching with accompanying increase in convenience, dependability, and saving of space. Control signals from the pushbutton array on the tape deck, from remote control, or from the fader relay are fed to an integrated circuit memory. Here they are stored and passed on to a decoder in an encoded form. Also fed into the memory is information from tape motion, direction, and end-of-tape sensors. The decoder translates all this stored information into logical tape deck operations, insuring that all tape movements are made properly and safely. This is a highly sophisticated and technical business, but the resulting operational advantages are of tremendous interest to all engaged in recording. The years ahead can only bring an increase in the application of such heady solid-state spinoff to the recording field. (We can hope for a decreasing price tag as well.)

SYNC OVERDUB

Sometimes a multitrack recording is acceptable except for one track. Maybe the musician on that track made a glaring mistake; perhaps he wasn't even present. Or maybe the director just changed his mind. Whatever the cause, it is necessary to record that one track or do it over to avoid the expense of calling the entire band back to repeat the number. Suppose the acoustic guitar on track 5 is the offender, and the other tracks are good. How do we solve this problem?

The delay between the record head stack and the playback head stack (resulting from the time it takes the tape to travel between the two) rules out the use of the playback heads as the cue tracks for overdubbing. If the playback head is used for cueing the musician, and the track is recorded at the record head, the overdubbed track will be out of sync all the way, too late by a fixed amount. The solution lies in overdubbing, which is made possible by switching of record heads over to *sync* position and using them temporarily as playback heads for cueing. In this way the track being recorded is in precise sync with the cue signals from the seven good tracks.

Referring to Fig. 5-4, we can see how sync overdubbing works. We thread up the 8-track tape from the recording session. We then throw switches to sync for tracks 1 through 4

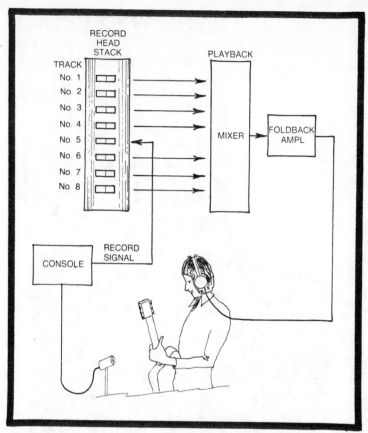

Fig. 5-4. Sync overdub is made possible by the use of recording heads as playback heads so that there is precise time synchronization with no delay between foldback (cue) signals and the track being recorded.

and 6, 7, and 8. These are routed through the board in a temporary mix to the foldback amplifier. The performer's headphones now allow him to hear the rest of the band. He can now play into a microphone which records only on track 5, the only one that needs to be redone. The erase head of track 5 removes the faulty track and record head 5 lays down a new one in its place in dead sync with the other 7. It makes no difference if the record heads are less than optimum in performance when used for playback—it's only for cue.

With this system it is possible to build up a multitrack recording bit by bit, instrument by instrument, or, for that matter, to make one person into a quartet by singing all the

Fig. 5-5. Forty tracks on 2-inch tape: the Stephens 40-track magnetic recorder/reproducer.

parts, one after the other. It is also possible to accommodate the schedules of expensive talent by shipping tapes across the country to add a few tracks of big-name performers. For example, in 1974 Apple Records released the album *Ringo*, enlisting the services of all former members of the renowned *Beatles* group. They recorded some of the tracks with Ringo Starr, John Lennon, and George Harrison in Los Angeles, and

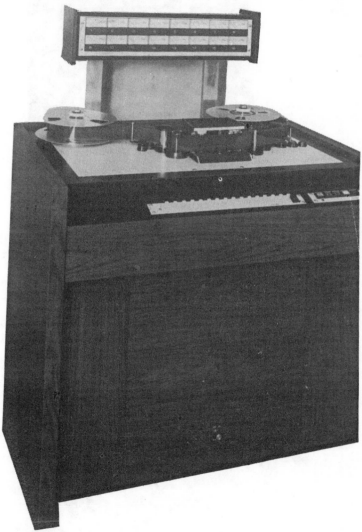

Fig. 5-6. The Scully 16-track magnetic recorder/reproducer using 2-inch tape, Model 100-16, manufactured by Scully-Metrotech.

Fig. 5-7. The Scully-Metrotech 8-track master magnetic re-corder/reproducer using 1-inch tape, Model 284-8.

then Mr. Richard Perry took the tapes to London, where he recorded Paul McCartney. Returning to L.A., the remaining tracks were recorded and the whole thing mixed down for release.

REPRESENTATIVE MULTITRACK RECORDERS

Figure 5-5 shows the imposing array of the function selecting knobs and VU meters of the Stephens 40-track professional recorder/reproducer. The sleek simplicity of the 2-inch tape drive is equally interesting. The VU meters of the recorder are more for lineup with the console, but reading 40 meters anywhere poses a problem in human engineering. Stephens offers a 32-track machine, recorders for 16- and 24-track work, as well as 8- and 16-track portable machines.

The Scully Model 100-16 magnetic recorder/reproducer is illustrated in Fig. 5-6. This recorder is without servo-controlled drive and without other sophisticated and expensive features, often duplicating what is on the console, yet it is a fully professional machine for 2-inch tape work. Scully claims a 50% reduction in cost by the elimination of redundant components and extra features. For example, a specially designed record/play head eliminates the need for separate playback head and overdub switching. The Scully Model 284-8, shown in Fig. 5-7, is a 1-inch, 8-track master machine which can be readily changed to work with ½-inch tape and a reduced number of tracks.

REFERENCES

1. Anon. *Digital Control For Sixteen Channels of Tape*. **db**. The Sound Engineering Magazine. Vol. 7, No. 2, February 1973 (pp 22—24, 26).

2. Soloff, Marvin. *Setting Up A Professional Recorder*. **db**. The Sound Engineering Magazine. Vol. 7, No. 2. February 1973 (pp 28—31).

3. Clive, Ross. *Electronic Signal Switching for Tape Recorders*. **db**. The Sound Engineering Magazine. Vol. 7, No. 3. March 1973 (pp 31—33).

4. Macdonald, Ross and C.A. Barlow, Jr. *Better Tape Head Azimuth Adjustments*. **db**. The Sound Engineering Magazine. Vol. 7, No. 3. March 1973 (pp 34, 36—37).

5. Karczmer, Claude. *A Modern Tape Recorder Design*. **db**. The Sound Engineering Magazine, Vol. 7, No. 3. March 1973 (pp 40—43).

6. Clemis, K. *Advanced Tape Mastering System: Mechanical Features*. Jour. Audio Engr. Soc., Vol. 12, No. 4, October 1964 (pp 303—307).

6 Monitoring Facilities

No one likes wavy window glass. Interposing something like that between the beautiful garden and our eyes is offensive to us. We also like clear, transparent audio chains. Perhaps the recording studio is more like taking a picture through that glass to pass on to others to enjoy. The sounds heard in the control room are to be passed on to others through recordings or other media. The human agents guiding this process must have a clean, clear, and undistorted sound at their ears to be able to do their job effectively. This is not simply a matter of having high quality loudspeakers or amplifiers, or even of having a good pair of ears, as important as these are. The room interposes itself between the loudspeaker and the operator and, unless proper precautions are taken, can have its own peculiar "wavy glass" effect.

Everyone working in sound recording must be aware of the human propensity to get used to a given type of sound, which is then equated with "good sound." A mixer who has worked for years in the old control room may unconsciously judge as inferior the acoustics of the new control room simply because they are different from the old one. Ears can be fooled, mistrained. Objective measurements and exposure to higher quality material can be important corrective steps.

MONITORING ROOM ACOUSTICS

Free intercutting of recordings made by different organizations or in various studios of a given organization is possible only if listening conditions match within certain rather narrow limits. Experience has shown that such matching rarely occurs without special effort. Of even more immediate importance than the intercutting of material from different studios is the moving of a given recorded tape from one control room (where it was recorded) to other rooms (for editing and mixdown). If the same tape sounds different to a given mixer when played back in different rooms, reference standards are lost and groping about in the darkness takes over.

It is amazing what strange antics a studio full of air is capable of, acoustically speaking. If we were to devise an equivalent electronic circuit of the studio space and analyze it in the audio band on the test bench, we would find a very complex circuit showing thousands of resonance points, very closely spaced in the higher frequencies but those below about 300 Hz spaced far enough apart to act singly and create problems. Any two reflecting parallel walls of an average sized control room will set up a standing wave at some audio frequency and at multiples of that frequency. A rectangular room with its six surfaces is capable of sustaining three *such* resonance systems, each having its own train of harmonics, and these constitute only the axial modes of vibration. The tangential and oblique modes complicate things even more but are, fortunately, of less practical significance. Even if walls are purposely made nonparallel, resonance effects are not eliminated—although they may be shifted and altered from those occurring in a rectangular room. If one is interested in a truly good listening environment, he must divest himself of the idea that the room acoustic problem is easily solved by good loudspeakers and a touch of equalization. Equalization cannot correct major acoustic faults, although it can minimize minor ones.

A detailed discussion of control room acoustics is outside the scope of this book, but those interested in looking further into the subject will find the material in Chapter 11 and the references helpful.[1-7] Here is a list of some of the more important factors to consider in building, renovating, or treating control rooms, as well as rooms used for editing and mixdown.

1. The minimum size of the monitoring room should be from 1500 to 4000 cu ft, discussed further in Chapter 11.
2. The basic dimensions of the room should adhere to certain proportions to distribute modal resonances most effectively. The following height/width/length ratios are preferred:[8]

H		W		L
1.00	:	1.14	:	1.39
1.00	:	1.28	:	1.54
1.00	:	1.60	:	2.33

3. Oblique surfaces such as inclined ceilings and splayed walls can be somewhat helpful in reducing flutter echoes, but these do not eliminate the standing-wave problem.

4. Concave surfaces tend to focus reflected sound and hence should be avoided. Double-glass observation windows should be approached with caution because of possible focusing effects.

5. The reverberation time (decay time) should be around 0.4 to 0.5 second and should be quite constant throughout the audible band.

6. Effort should be made to distribute the various types of absorbing material among the different surfaces so that it might be effective on the various modes of vibration.

7. The platform on which the console and recorders often rest can be a valuable active element in the acoustic treatment of the room if low-frequency absorption is needed. Through diaphragm action it will absorb sound in the low-frequency region; with flooring of ¾ in. plywood and a 12 in. airspace it will peak in the 30−40 Hz region, and its effect may be broadened but not nullified by the usual practice of covering with heavy carpet. This low-frequency absorption may be helpful in controlling vertical normal-mode resonances. If low-frequency absorption is not needed, the floor of the platform can be made of 2 in. tongue-and-groove decking lumber.

8. The rear wall is normally made acoustically highly absorptive over the full band and, in addition, bass absorbers are normally required to control the inevitable isolated and grouped modal resonances which occur in spite of the best efforts toward distributing them through room proportioning. The side walls may provide some reflection.

9. After the room is acoustically treated the room response should be determined and proper compensating equalization inserted before the power amplifiers. Uniform listening conditions should prevail over the area occupied by mixer, producer, and others directly involved in sound evaluation. Electrical equalization should be used only for trimming—not for correction of major acoustic flaws.

10. Because of the desirability of balance between loudspeakers in 2- and 4-channel stereo work, there should be bilateral symmetry in the placement of loudspeakers, console, and other control room

furnishings as far as possible. This applies to acoustical treatment as well, because acoustical balance is the goal. With glass windows in front of the mixer, there is a departure from the ideal of symmetry affecting balance of the rear loudspeakers with the front ones. This may be corrected electrically.

THE HOME ENVIRONMENT

Disc and tape recordings, radio, and television all find their chief ultimate consumer in the home. Radio and tape are used commonly in automobiles, but there is little point in straining over quality in this far-from-quiet environment. The home does have reasonable quality potential and it is well for the producer of records to consider the ultimate home environment in which his records will be played. What goes on in the monitoring room should be related to home listening conditions. Should a 6 in. loudspeaker be used in the TV control room because that is what is used in the average TV receiver? Records are often reproduced on small players having loudspeakers not much better. There is an obligation, encouraged by competitive factors, to offer the consumer the best quality possible so that he may be encouraged to strive for better quality in his reproducing system. The key is in the hand of the record producer, for if only records of poor quality are available, the industry stagnates and quality goals languish everywhere.

If the control room reverberation time is comparable to that of the average living room, the recording engineer can make his judgments with the assurance that, on the average, what he is hearing is close to that which the customers with better equipment, at least, will hear. At least two investigations of living room acoustics have been conducted in England, the first in the homes of 16 BBC engineers[2] and the second covering 50 average homes.[9] Because construction and furnishings have such a strong influence on reverberation time, there is some question as to how applicable such measurements are to U.S. homes. They do, however, serve as a guide and we note that average values of reverberation time of 0.5−0.7 second were found in the 200 Hz region, and about 0.4 second at 5 kHz. This is near what has been recommended for our control room.[3]

EARLY SOUND AND OVERHEAD REFLECTORS

Those having extensive practical experience in recording make frequent reference to *early reflections* as a very

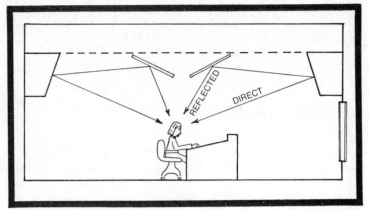

Fig. 6-1. Overhead reflectors result in constructive and destructive interference between direct and reflected rays.

desirable attribute in the design of a monitoring room. The term is generally reserved for use in the field of auditorium acoustics and the recent application of *clouds* (reflecting surfaces suspended over the audience) is to get reflected sound from a source on the stage to the ears of the listener over the shortest possible path. A reflection does not sound like an echo unless it arrives more than about 50 msec after the direct sound. This is called the *precedence* effect. Any sound arriving within about 35 msec is perceived by the human ear as direct sound.

Let us consider this with the mix engineer at his board in the control room in mind. As sound travels about 1.1 ft/msec, the reflected sound would have to travel more than 50 ft after the time of arrival of the direct sound at the ear of the listener to be perceived as reverberation. If the average control room has dimensions of the order of 25 ft, and the mixer is not far from the center of the room, at least first reflections from walls and possibly even second and third reflections will arrive at the mixer's ears within 35–50 msec. With the usual absorbent conditions, there may be a significant loss at each reflection in addition to the normal falloff with distance. Therefore, there seems to be little support for bringing the *early sound* concept into consideration of monitoring rooms in justification of the use of reflecting surfaces over the mixer's head.

In the geometry of Fig. 6-1 the direct ray reaches the mix engineer's ear a certain time after leaving the loudspeaker. The reflected ray arrives about 2 msec later in the usual monitoring room arrangement. This delay amounts to a phase shift in degrees depending upon the fraction of a wavelength or the number of wavelengths that the 2 msec figure represents. A delay of 2 msec represents a full wavelength (360 or 0°) at 500 Hz, a half wavelength (180°) at 250 Hz. Down through the audio band the ray delayed 2 msec pulls in and out of phase with the direct ray as they combine at the mixer's ear. The presence of the reflecting surface approximately doubles the sound pressure at the console (6 dB increase) when the two components are in phase; when they are out of phase, deep notches in response occur. As the constructive interference regions are broader, the general effect of the reflecting surface is to add the 6 dB in sound pressure level plus a few dips. The nulls tend to be blurred by surface irregularities, especially in the mid and high frequencies, but can be pronounced in the lows.

A similar situation with respect to microphones has been carefully investigated by Anderson and Schulein.[10] A 10 dB dip at 230 Hz was measured with the microphone 53 in. above the floor and the source 15 ft away, the difference in path length between the direct and floor reflected ray being about 1 msec. Response irregularities of the order of 5 dB were also noted with a different geometry, giving a delay of about 4 msec.

There would seem to be some question about the desirability of using such reflecting surfaces in a monitoring situation. Extreme efforts are made to achieve flat response across the band and across the console—and then the equivalent of a comb filter is employed in the form of such reflecting surfaces. This introduces a mild case of flanging or phasing into the program material (see Chapter 7). Here are some arguments in defense of reflectors:

- There is a desirable subjective effect of some kind that offsets the disadvantages.
- The major dips can be equalized electrically;
- A twofold increase in level is needed.
- The reflectors are nice to hide bass traps above (or to impress clients!).
- The reflectors favor the highs which is good in a small room that emphasizes the lows.
- The reflectors add a *bright* sound.

The frequencies at which these peaks and dips occur, as well as their magnitudes, are determined by the physical arrangement, reflector surface conditions, etc. A rough reflector surface would tend to smear the nulls in the high frequencies, but would have little effect at the mid and low frequencies. Because the dips are fixed, they can be equalized electrically, but that is about all that can be done with them.

This procedure violates our premise of using equalization only to correct unavoidable irregularities.

When spaced microphones are used for stereo pickup, source localization is conveyed by a combination of intensity and time of arrival differences between the two signals. In stereo the differences in time of arrival do not result in a comb filter effect. If there is appreciable overlap in the pickup of the two microphones, however, the comb filter effect rears its ugly head when the stereo record is played back in mono. Reflectors in the monitoring room may be better tolerated in stereo monitoring, but in checking for mono compatibility the irregularities in response due to the comb filter effect would be there to trouble the operator.

MONITORING LOUDSPEAKERS AND AMPLIFIERS

It is generally agreed that the weakest link in the audio chain is the loudspeaker. Microphones are better, amplifiers are better, but reproducing sound waves in air which follow closely the original sound waves falling on the microphone is inherently a difficult problem aggravated by the nine- or ten-octave span of the human ear and the fact that any practical radiating element is very small in terms of wavelength of sound at the bass end of the audible spectrum.

While tremendous strides have been made recently in loudspeaker refinement, few revolutionary innovations have really caught on—we find the electrodynamic type of driver, ancient but highly refined, still dominating the scene, with electrostatic elements popping up here and there. We should not minimize the great improvements which have been and are now being made in electrodynamic loudspeakers. New materials have made stronger magnets possible and voice coils of greater diameter and width execute greater excursions with less distortion than ever before.

The loudspeaker and the power amplifier driving it should be considered together. The performance of the amplifier is affected by its load (loudspeaker), and the load the loudspeaker presents to the amplifier is affected by the frequency, the characteristics of the voice coil, the design of the cabinet, and even the room. Many of these problems are minimized by the new solid-state amplifiers and their very low internal impedance.

A given electrodynamic loudspeaker element works well over only a portion of the audio spectrum. Separate elements specifically designed for efficient operation in a portion of the frequency range give superior results. Studio monitoring loudspeakers today are two- to four-way devices. A three-way

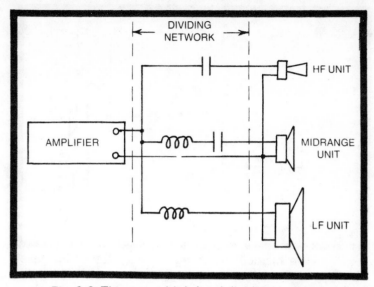

Fig. 6-2. Three-way high-level dividing network.

dividing network, shown in Fig. 6-2, divides the audible spectrum between the three units more or less as shown by the curves of Fig. 6-3. In this case the crossover frequencies are 800 Hz and 5 kHz but there is great variability in crossover frequencies used in commercial practice. The crossover point is the half-power point on each curve (3 dB down) so that energy of one band adds to that of the adjoining band, giving a

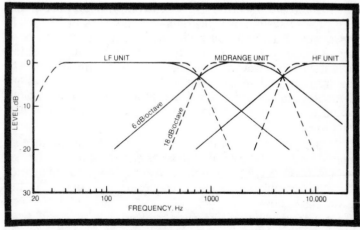

Fig. 6-3. Crossover points of typical three-way loudspeaker system. Crossover frequencies are not standardized.

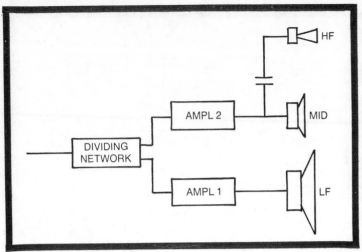

Fig. 6-4. The biamplifier arrangement utilizes a low-level dividing network except for the high-frequency unit.

smooth transition from one unit to the other. The crossover network of Fig. 6-2 has skirts falling off at 6 dB per octave, although they are often much steeper. The ultimate low-frequency limit is determined by the falloff of the response of the low frequency woofer. Even though modern solid-state audio power amplifiers have good response to 50 or 100 kHz, the tweeter invariably limits the high-end response.

Biamplifiers and Triamplifiers

The crossover network is an extremely vital part of the loudspeaker system, and the high-level network of Fig. 6-2 is the source of a number of problems. The dividing network of Fig. 6-4 operates at low power level with many resulting advantages. However, two audio power amplifiers are required instead of one, hence the name biamplification. Figure 6-5 illustrates a triamplifier arrangement. Even 4-band amplification is becoming common in high power sound reinforcement.[11-14]

Although two or three amplifiers would appear to cost more than one, the overall economics can work out in favor of the low-level network system. Certainly the performance of the multiamplifier system is superior, as the following advantages show:

- Intermodulation distortion reduced.
- Overload distortion reduced.
- Transient response significantly improved.

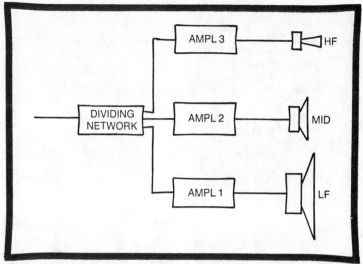

Fig. 6-5. The triamplifier arrangement uses a separate power amplifier for each loudspeaker element.

- Much more flexibility in handling crossover network design with resulting improvement in quality and steepening of filter skirts (see the 18 dB/octave curves in Fig. 6-3).
- More effective use of audio power and increase of dynamic range by using a plurality of smaller power amplifiers.

Studio Monitors

As examples of quality loudspeakers widely used in recording studios, three models of Altec studio monitors are shown in Fig. 6-6. Altec engineers believe that the fewer crossover points the better; accordingly, these are all two-way loudspeakers which may be used with single amplifiers and high-level crossover networks or with biamplifiers and low-level networks. Sealed enclosures, high-compliance surrounds, and large magnet structures are utilized to give good transient response. The high frequencies are handled by compression drivers mounted on sectoral horns which give typical dispersion of 40° vertically and 90° or 100° horizontally.

Figure 6-6A illustrates the smaller Model 9849A, which weighs 60 lb and has a rated power capacity of 60W. It has a specified frequency response of 40—15,000 Hz and a 900 Hz crossover with 12 dB/octave slope. This unit produces a sound pressure level of 93 dB at 4 ft with 1 watt of pink noise at

Fig. 6-6. Altec studio monitors: (A), Model 9849A, rated at a power capacity of 60 watts; (B), Model 9846-8A, rated at 100 watts; and (C), Model 9848A, rated at 200 watts.

operating bandwidth. Normal program material has many peaks which must be handled by the monitor loudspeaker without distortion. The 60W rating is RMS and it is common to leave 10 dB of "headroom" for peaks—which means that the complex program power level would be only 6W.[5] To obtain an estimate of the sound pressure level at the mixer's ear 8 ft from the loudspeaker, we must subtract 6 dB from 93 dB for doubling the distance from 4 to 8 ft and add 7.8 dB because we have 6W rather than 1W input. This gives a working sound pressure level at the mixer's ear of about 94.8 dB.

Altec's Model 9846-8A monitoring loudspeaker, pictured in Fig. 6-6B, is the next larger unit—with a power rating of 100W and a stated frequency response of 30 to 15,000 Hz. Its output, at 1W input, is 91.5 dB at 4 ft. Leaving 10 dB headroom again (10W operating level) and going out to 8 ft, this monitor will produce a sound pressure level of 95.5 dB.

The larger monitor in this series is the Model 9848A shown in Fig. 6-6C. This unit features an output of 96 dB/4 ft/1 watt, and a 20−15,000 Hz frequency response. With 10 dB headroom (10W input) and for a position 8 ft from the loudspeaker, the output power produced by this unit is 103 dB. (The pain threshold in human hearing lies in the 100−120 dB range.)

A few decibels of gain is purchased at great increase in size and cost. If higher levels than these are required, a compromise might be made with the 10 dB headroom figure, or that 6 dB gain by using overhead reflectors might look pretty good even with a few notches in the response.

EQUALIZATION OF MONITORING ROOM

Modern music is made in the control room, it is said. In fact, it is often made in many control rooms as several studios may be involved in assembling different tracks which go to make up the finished product. Acoustical consistency between different control rooms and between editing and remix rooms of a single organization means that a mixer can play his recording in different rooms and still recognize the final product as his own. It is obvious that for a given recording to sound the same to a given person, even though listened to in different rooms, would be a very valuable feature in a large studio complex. Equalization of such rooms is the accepted method of bringing them to a degree of standardization.[15, 16, 17] First, let's review just what is meant by equalization of control rooms.

Resorting to equalization techniques to achieve satisfactory listening conditions underscores the very great influence of the acoustics of the listening space on the quality of the sounds heard. It has just been stated that the weakest link in the audio chain is the loudspeaker. We must now revise this to say that the weakest link in the audio chain is the loudspeaker and the space in which it operates, taken together. The pretty response curves of a loudspeaker obtained in an anechoic test room may have little similarity to the real-life response of that loudspeaker in a monitoring room. It is to make the monitoring loudspeakers and the control room work harmoniously together that equalization procedures are widely applied.

It is a serious mistake to look to equalization to correct for poor acoustic design of the monitoring room. As has been emphasized previously, the great bane of small studios and control rooms—say, less than 5000 cu ft volume—is normal-mode resonances and resulting standing-wave patterns. If "holes" and "hot spots" and differences in frequency response are noticed as one's listening position is shifted about the room, they will still be there after the room is equalized, as long as the sound sources remain fixed. Equalization can, however, correct for broad deviations from ideal acoustic conditions in a room as well as the broader variations in loudspeaker response. It cannot cure a bad listening situation but it can make an acceptable one excellent.

Just what is room equalization? We are interested in sounds as sensations experienced by producers, directors, and mixers. This involves complicated physiological and psychological processes, the depths of which we cannot probe in this book. The loudness of a sound, the perception of pitch and timbre, all involve the so-called critical bands of the human ear. With great oversimplification we can say that our ears act like a set of some 24 bandpass filters having center frequencies from 50 Hz to 13.5 kHz and bandwidths as shown on the critical band curve of Fig. 6-7. For example, a critical band centered on 700 Hz has a bandwidth of about 140 Hz. Each band corresponds to a distance of 1.3 mm along the basilar membrane of the inner ear. These bands operate quite independently of each other. For example, a pure tone falling in one critical band is not readily masked by other components outside this band. In other words, the ear is constantly performing real-time frequency analysis of sounds falling on it and it is well for us to have at least a nodding acquaintance with such an awesome natural mechanism upon which we lean so heavily in audio work.

Ideally, room equalization measurements should be made using filters having the same passband as the critical bands of

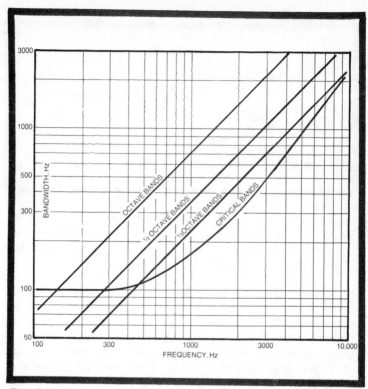

Fig. 6-7. Critical bands of the human ear compared to octave, half-octave, and one-third-octave bands of filters.

the ear. In Fig. 6-7 we note that bands one-third octave wide are only a rough approximation to critical bands, but, because they are an approximation and are readily available, they are commonly used in room equalization procedures.

The first step in equalizing a control room is to make sure that the basic acoustical design of the room is right, that all acoustic treatment is installed, and the performance is verified by objective measurements. The second step is to obtain the acoustic response of the room. This is done by playing 1/3 octave bands of *pink noise** through the loudspeakers (taken one at a time) by injecting the signal at the power amplifier input and adjusting to a constant level at this point. The sound from the loudspeaker is then measured

---

*Random noise is called *white noise* because the energy is uniformly distributed throughout the audible spectrum. comparable to the energy distribution in white light. *Pink noise* drops off 3 dB per octabe with increasing frequency so that the energy in each 1/3 octave band will be the same.

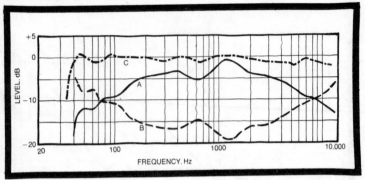

Fig. 6-8. Equalizing the acoustic response of a hypothetical control room: (A) room curve as determined by measurements, (B) inverse of room curve giving equalization required, and (C) the resulting overall response at the point where the (A) measurements were made.

by a sound level meter whose microphone is placed at the chosen listening spot—perhaps where the mixer's ears are normally located. By plotting the sound level meter readings against the 1/3 octave band center frequencies, the *house curve*—acoustic response for that particular room, position, amplifier, and loudspeaker—is obtained as in curve A of Fig. 6-8: We note considerable dropping off at each end of the spectrum, but are relieved to see no wild peaks or valleys between.

The third step is to equalize the deviations of curve A. The complementary compensation required is sketched in curve B. A network with the curve B response is then inserted in the chain before the power amplifier, which should yield the overall response depicted as curve C. This network may be calculated and built up on a fixed basis; as an expensive alternative, an adjustable 1/3 octave equalizer may be inserted and adjusted to give the proper compensation. The equalizer now brings the power amplifier/loudspeaker/room combination into a reasonably flat condition for that particular loudspeaker and listening position.

Most of the compensation is that required by the loudspeaker and room because amplifiers are normally quite flat. Changing the loudspeaker or the acoustic treatment of the room upsets everything, and it's back to step one or step two. Accidental overload on loudspeakers, such as might occur in fast rewind of a tape recorder, can cause shifts in their response. Touching up the equalization periodically can compensate for these if the loudspeakers are not damaged.

Magnitude of Room Effect

How much of the irregularity of acoustic response of the room of Fig. 6-8 is due to the loudspeaker and how much to the room? An interesting comparison has been presented by Eargle and Engebretson, which illuminates this question.[5] Figure 6-9 shows acoustic response graphs for identical left and right loudspeakers (Altec Model 9846B), but mounted in

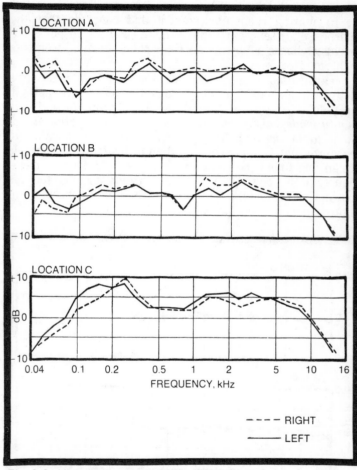

Fig 6-9. Unequalized acoustic response of three different rooms[5], each using a pair of Altec 9846B loudspeakers: (A) typically good control room; (B) control room asymmetrical, cavity responses between and behind loudspeakers; (C) remix facility, heavy carpet and draperies, wallboard construction, large windows.

three different rooms. Location A is a typical, good control room. Location B is an asymmetrical control room where the loudspeakers are deployed without proper baffling and with consequent cavity resonances between and behind the loudspeakers. Location C is a remix facility of unusual characteristics. Heavy carpet and draperies provide an overabundance of absorption in the high frequencies while wallboard construction and large windows give considerable absorption in the lows. We note the very significant effect of the monitoring room environment and conclude that the loudspeaker itself must have quite a flat response, a conclusion borne out by the anechoic-chamber measurements on this loudspeaker.

It turned out that location A did not need equalization, a fact which tends to bear out the contention of some experts in the field: *If the room is really designed properly and high quality loudspeakers mounted correctly, equalization is not likely to be needed.* The problem comes in learning to recognize that you have achieved an acoustically acceptable design and found the right spot for the right loudspeakers. Figures 6-10 shows the after-equalization graphs for the control room B and the remix room C and the resulting great improvement in overall response.

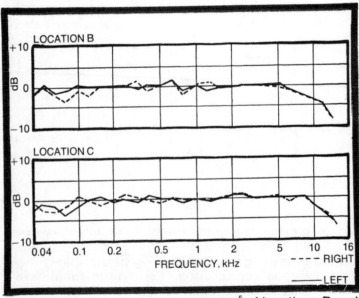

Fig. 6-10. Equalized acoustic response[5] of locations B and C of Figure 6-9. Location A required no equalization.

Should the High End Roll Off?

Many monitoring systems of recording organizations roll off gently above 5 to 7 kHz. Is this intentional or is it just the way things worked out? Surely, as the frequency is increased the loudspeaker pumps less power into the control room because of several factors, not the least of which is the increased directivity. The omnidirectional microphone of the sound level meter used in obtaining the acoustic response of the room would tend to fall off in the highs as it indicates, more or less, the total acoustic power present in the room. We must also remember that the human hearing apparatus has certain directional tendencies. It would seem logical and otherwise desirable to equalize to a flat condition and, if deviations are needed, to introduce them specifically. After all, the average modern recording studio probably has several dozen equalizers available and others can be switched in to meet specific needs. Without a flat system base to work from, the entire process of program equalization tends to disintegrate.

What is the Best Listening Distance?

How close should one be to the loudspeaker for most critical listening? Don Davis has discussed this point with recording studio monitoring specifically in mind.[16] He makes a clear distinction between the direct sound and the reverberant field, and, of course, in the studio there are areas where they cannot be separated except by sophisticated pulse techniques. The direct sound level falls off at a rate of 6 dB per doubling of the distance (Fig. 6-11) except for (a) the first 4 or 5 ft (the so-called near field) where measurements are erratic, and (b) the greater distances complicated by reflected energy. At a certain critical distance the direct and reverberant fields are equal. Beyond the critical distance the reverberant field of the control room dominates, and this tends to remain at a relatively constant level everywhere in the room except near the loudspeakers. Davis suggests, and those in the field seem to agree, that the most critical listening can be done closer to the loudspeaker than the critical distance, but not so close as to get in the erratic near field. In this preferred region the sound coming over the loudspeakers can be evaluated for its own reverberant character while at greater distances the control roon reverberant field would tend to dominate the signal.

The size of the control room and the amount of absorption in it determines the level of the reverberant field. A very live acoustic would tend to reduce the preferred listening region, a

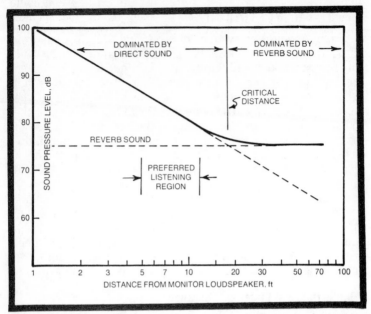

Fig. 6-11. Relationship of the preferred distance between operator and studio monitor to critical distance of the room.

dead acoustic to increase it. From this perspective we can see more clearly the importance of having reverberation time constant throughout the audible band to avoid response shifts with time. We also see the conflict between having constant level over an appreciable area and proper ratio of direct to reverberant field.

Is it proper to say, "Eight feet is the best distance for the most critical listening?" We must distinguish between physical distance and acoustic distance. In one control room eight feet might put you well into the region dominated by reverberant energy. There is, however, logic in comparing two listening situations on the basis of acoustic distance.

Recording studios today employ 2, 3, or 4 monitoring loudspeakers. If these are energized by stereo or quadraphonic signals, the preferred listening area tends to be reduced because of the need to keep relative balance within limits.

HEADPHONE MONITORING

We may be coming into an era in which headphone monitoring could become very attractive because of recent developments in headphone quality. There are many situations in which high ambient noise levels and other distractions could make headphone monitoring far superior to loudspeaker monitoring. The control room is often the scene of extraneous activities—at least as far as the present recording job is concerned—and headphones might rescue a bad situation.

Recent developments in electrostatic headphones have extended the usual five octaves to nine, which is wider than our loudspeaker systems.[18] In that cruel 400 Hz square wave test, these headphones far outperform loudspeakers. This will be an interesting field to watch as measurement techniques are refined so that improvements in design can be adequately evaluated. It is very difficult to make meaningful headphone measurements that correlate with what the ear hears.

Quadraphonic headphones burst full bloom upon the world overnight,[19] at least as far as the consumer market was concerned. There is much discussion as to the magnitude, quality, and even the existence of quadraphonic effect in these headphones. At this time they stir little excitement in the professional recording field.

REFERENCES

1. Queen, Daniel. *Monitoring Room Acoustics*. **db**. The Sound Engineering Magazine, Vol. 7, No. 5, May 1973 (pp 24—26).

2. Gilford, C. *The Acoustic Design of Talk Studios and Listening Rooms*. Proc. IEEE, Vol. 106, Part B, No. 27, May 1959 (pp 245—256).

3. Gilford, Christopher. *Acoustics For Radio and Television Studios*. IEEE, Monograph Series 11, Peter Peregrinus, Ltd., 1972 (pp 221—225).

4. Everest, Alton. *Acoustic Techniques For Home and Studio*. TAB Books, 1973 (Chapters 4, 5, an 6).

5. Eargle, John and Mark Engebretson. *A Survey of Recording Studio Monitoring Problems*, Recording Engineer/Producer, Vol. 4, No. 3, May/June 1973 (p19).

6. Veale, Edward. *The Environmental Design of a Studio Control Room*, presented at the 44th Audio Engr. Soc. Convention, Rotterdam, February 1973 (Preprint A-2R).

7. Davis, Gary. *Control Room Acoustics*, Recording Engineer/Producer, Vol. 5, No. 2, March/April 1974 (p31).

8. Sepmeyer, L. *Computed Frequency and Angular Distribution of the Normal Modes of Vibration in Rectangular Rooms*, Jour. Acoust. Soc. Am., Vol. 37, No. 3, March 1965 (pp 413—423).

9. Jackson, G. and H. Leventhall. *The Acoustics of Domestic Rooms*, Applied Acoustics (5) 1972.

10. Anderson, Roger and Robert Schulein. *A Distant Micing Technique*. **db**. The Sound Engineering Magazine, Vol. 5, No. 4, April 1971 (pp 29—31).

11. Siniscal, Albert. *Bi- and Tri-Amplification*. Recording Engineer/Producer, Vol. 2, No. 2, March/April 1971 (pp 27—29).

12. Siniscal. Albert. *High Intensity. Modular Tri- and Quad-Amplification/Loudspeaker Systems.* presented at the 42nd Convention of Audio Engr. Soc.. May 1972 (Preprint 868 K-7).

13. Ashley. J. and Allan L. Kaminsky. *Active and Passive Filter Networks as Loudspeaker Crossover Networks.* Jour. Audio Engr. Soc.. Vol. 19. No. 6. June 1971 (pp 494—501).

14. Smith. Allan. *Electronic Crossover Networks and Their Contribution to Loudspeaker Transient Response.* Jour. Audio Engr. Soc.. Vol. 19. No. 8. September 1971 (pp 674—679).

15. Eargle. John. *Equalizing the Monitoring Environment.* Jour. Audio Eng. Soc.. Vol. 21. No. 2. March 1973 (pp 103—107).

16. Davis. Don. *Considerations in Acousta-Voicing Studio Monitors.* **db**. The Sound Engineering Magazine. Vol. 4. No. 12. December 1970 (pp 26—27).

17. Flickinger. Daniel. *Electronic Adjustment of Monitoring Acoustics.* Jour. Audio Engr. Soc.. Vol. 18. No. 6. December 1970 (pp 657—661).

18. Souther. Howard. *Improved Monitoring With Headphones.* **db**. The Sound Engineering Magazine. Vol. 3. No. 3. February 1969: and Vol. 3. No. 4. March 1969 (pp 28—29).

19. Test report. *Quadraphonic Headphones.* Audio. Vol. 57. No. 6. June 1973 (pp 28—31).

Special Effects 7

The great advantage of separation recording is the freedom it gives to alter and adjust the original in the mixdown. It gives freedom to adjust the balance between the outputs of the various musicians of a musical group, freedom to touch up frequency response of any channel, freedom to move any input to any output bus in mono, stereo, or quad, freedom to fade down one performer during a flub and thus salvage a take, and freedom to delete or add tracks at will by overdubbing. In many forms of contemporary recording, perhaps the most important freedom is that of introducing special effects of many kinds and in any desired degree.

REVERBERATION

We have discussed reverberation (echo) in the context of space-planning for reverberation chambers or equipment, the various devices for generating artificial reverberation, and the routing of reverberation signals on the console. There isn't too much more to be said other than reverberation must be considered a special effect, and one of the more important ones in adding spatiality to the final recording. This is needed with separation recording more than ever before because of the use of close microphones and soft studios. During the recording there is extensive use of temporary mixes with reverberation added. In fact, it is quite common to record these temporary mixes on a spare recorder for reference use. It must be emphasized, however, that the original tracks must be recorded dry (without reverberation) if the fullest freedom of manipulation is to be retained, although, under pressures of schedule and economics, shortcuts must often be taken.

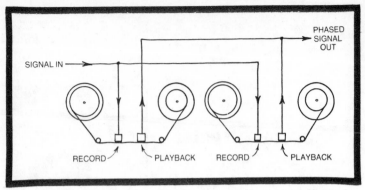

Fig. 7-1. The generation of phasing or flanging effects by use of two tape recorders.

PHASING

There is tremendous activity in popular music today to produce sounds more weird and far out than the competition. The genius expended in this direction is as awesome as the sounds produced. Historically, one of the earliest techniques of this type was flanging or phasing. It was first produced with two tape recorders arranged as in Fig. 7-1. A signal was recorded on two tape recorders simultaneously and the playback signals from both machines combined. By changing the speed of one recorder slightly with respect to the other, a whirling, swishing, space-like, inside-out effect can be created. The speed change may be accomplished by lightly pinching the capstan, first of one machine and then the other, or applying some pressure to the flange of the supply reels, the source of the term *flanging*. Slowing one machine with respect to the other introduces a time delay, so some frequencies will be exactly out of phase with the corresponding frequencies of the other machine, resulting in selective cancellation. Other frequencies will be in phase, resulting in constructive combination. Of course, the phased output in the arrangement of Fig. 7-1 is delayed with respect to the original signal by the time it takes the tape to travel from the record to the playback head and hence would not be suitable for live performances. There are innumerable ways the amateur can combine signals from discs and multiple tape recorders to achieve this phasing effect.[1] All that is required is that two sources of the same signal are combined at similar amplitudes with provision for introducing a time difference between the two.

What we actually have in phasing is a variable comb filter, as Bartlett pointed out.[2] When two signals are combined with a given delay between them, the effect of the delay depends upon the frequency. For example, a delay of 1 msec represents a phase shift of 360° at 1 kHz, but only 180° at 500 Hz. A shift of 360° means the two signals are in phase, so they will add, giving a 6 dB increase in output voltage. A phase shift of 180° places the two signals in phase opposition and they cancel. The resultant of combining two signals of equal amplitude with a delay of 0.4 msec between them is shown in Fig. 7-2. Because 0.4 msec is a full wavelength at 2.5 kHz, we see in Fig. 7-2 a peak at this frequency, and for each integral multiple of 2.5 kHz (5, 7.5, 10, 12.5, and 15 kHz) and nulls midway between. For delays less than 0.4 msec there would be fewer notches in the audible band; for greater delays more notches would result. Thus we see that phasing or flanging introduces this variable comb filter phenomenon, which yields its characteristic phychedelic effect, a resource for the creative effects man.

Phasing By Phase Shifter

A manually controlled phase shifter allows the introduction of phasing or flanging effects in live presentations and in situations in which the effect is required instantaneously without messing around with several tape recorders. In Fig. 7-3 a dividing network and a combining network have been added to the phase shifter to avoid circuit

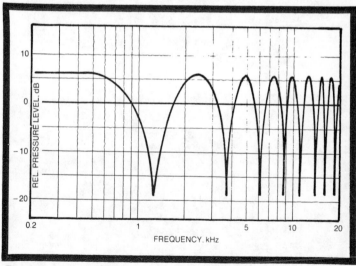

Fig. 7-2. The comb filter effect of phasing when two signals of equal amplitudes are combined with a delay between them of 0.4 msec.

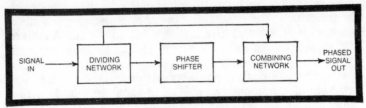

Fig. 7-3. Schematic for phasing by phase shifter.

interaction, but otherwise it is the equivalent of the two-transport method of Fig. 7-1.

Phase shifters marketed by Countryman Associates, which are well adapted to flanging, are shown in Fig. 7-4. The units are designed to process signals at standard studio line levels with unity gain; they are equipped with internal dividing and combining networks. It is possible to achieve a wide range

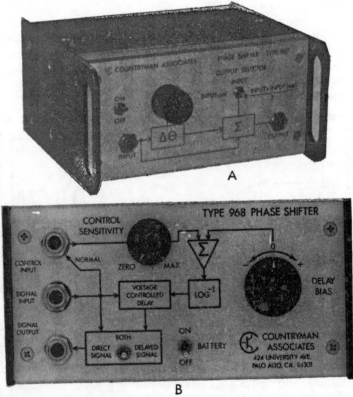

Fig. 7-4. Countryman Associates Types 967 and 968 phase shifters.

144

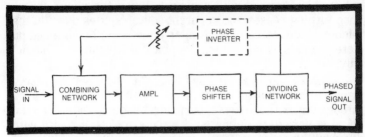

Fig. 7-5. More melodic flanging effects are obtained by feeding back some of the delayed output. (Countryman Assoc.)

of phasing effects with these devices.[3] For example, Fig. 7-5 diagrams a method for obtaining more melodic reel flanging effects with less deep cancellation frequencies. This is accomplished by feeding some of the delayed output of the mixed output back into the input either in or out of phase. Feeding back the mixed output gives a more pronounced phasing effect and greater control through the feedback control.

Two- and Four-Channel Stereo Phasing

Stereophonic motion effects can be obtained from monaural signals with the hookup of Fig. 7-6. The human hearing mechanism is normally quite insensitive to phase differences as compared to relative amplitudes; and if the phase shifter control is not moved, there is no apparent subjective effect. When the phase shifter control is changed, however, the ear notes the effect and the source seems to shift across the stereo field. As different parts of the frequency domain are affected differently, as shown in Fig. 7-2, the apparent source of the various frequency components will shift in different ways, giving a blurred motion effect quite

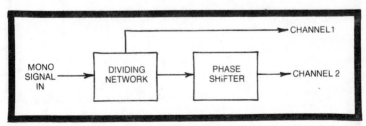

Fig. 7-6. Stereo motion effects may be generated from monaural signals. By sending the output through a matrix quadraphonic decoder, 4-channel effects may be obtained. (Countryman Assoc.)

unlike that obtained with a panpot. When the phase shifter knob turning ceases, the apparent source settles down in the middle of the stereo field. Similar effects can be obtained on stereo signals by connecting the phase shifter in one channel only.

A stereo signal derived as in Fig. 7-6—or a normal stereo signal with a phase shifter in one channel—can be played through any of the matrix-type quadraphonic decoders. As the delay is varied, the apparent source circles the room, with different frequency components moving in different directions. Placing a phase shifter in each of any two adjacent channels of discrete four-channel stereo can yield effects similar to the stereo effects above if the controls of both shifters are moved together. If only one channel is delayed, there is no dramatic effect.

An adaptation of the stereo setup of Fig. 7-6 is shown in Fig. 7-7. The stereo signal output is developed from the monaural channel. Channel 1 is driven by the in-phase flanging mix and channel 2 by the out-of-phase flanging mix. The interesting effect here is that frequencies which cancel in

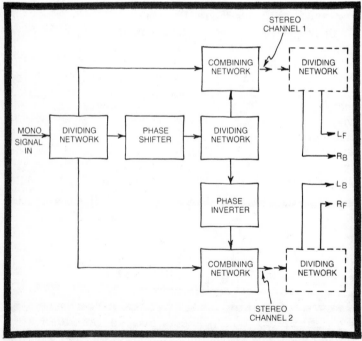

Fig. 7-7. Alternate approach to generating quadraphonic flanging effects from a mono signal. (Countryman Assoc.)

channel 1 appear in channel 2 and vice versa. Different frequency components seem to come from different directions; by varying the phase shifter, the apparent source of the different frequency components moves across the stereo field along with the changing flanging effects. Playing this stereo signal through a matrix-type quadraphonic decoder causes the sound to circle the room, but now the apparent sources of different frequency components will have different relative positions and may move in different directions.

Spatial Distortion

Discrete quadraphonic effects can be obtained, as mentioned earlier, simply by driving left front and left back channels with one stereo channel and right front and right back with the other. If the arrangement of Fig. 7-7 is fed to four discrete channels in a diagonal pairing arrangement as indicated, even more interesting sounds may be produced. Flanging effects are heard on all four channels, but frequencies canceled in one diagonal pair appear in the other. If the phase shifter control is now varied, the result is an apparent stretching or squeezing of space along the two diagonal axes, different frequencies being stretched or squeezed on different axes at different times, producing a most unusual effect. Signals rich in harmonic content, such as a harmonica passage, show the effect best.

Spatial and other flanging effects can also be controlled by components of the musical signal. For example, a bass drum signal can be made to cause the snare drum or cymbal sound to go through a flanging sweep every time the bass drum is struck. There is practically no limit to the unusual effects an inventive person can achieve through the methods outlined above and elsewhere. In Fig. 7-8, the Eventide *Instant Phaser* is shown. The period of the phasing can be controlled by means of the built-in oscillator. By using the envelope follower, phasing can be controlled by the amplitude of the incoming signal as described above.

Digital Audio Delay System

The application of digital techniques to signal processing opens up a Pandora's box of bright new possibilities. Until

Fig. 7-8. The Instant Phaser by Eventide.

recently, linear technology held sway in amplification, transmission, and processing of audio signals. Voltages or currents of various waveforms have always done the job for us; so why revert to pulses which seem so far removed from our nice, clean audio signals? The answer to this is simply that in many ways a better job can be done digitally than linearly: better signal-to-noise ratios, less distortion, and the ability to do many things which just cannot be done otherwise, or to do them better and cheaper. An audio signal can be delayed by various tape systems, but all the problems associated with magnetic tape and electromechanical systems are the inevitable accompaniment. Digital techniques have made possible an all-electronic device which is vastly superior to tape systems.[4,5]

To enter into the digital world we must convert our familiar analog form of signals as voltages varying with time to the binary form. The familiar signal form shown in Fig. 7-9A is the analog form. This signal may be *sampled* at short intervals by the timed pulses of Fig. 7-9B. The sampling process is, in effect, a multiplication of the audio amplitude at that instant by the sampling pulse which has an amplitude of unity, yielding the audio envelope signal of Fig. 7-9C. Right here the newcomer to the digital field wonders how his precious audio signal waveform sampled only now and then can ever be put back together for the faithful rendition of the original. This requires either a little faith or much study, but it can be done with essentially perfect fidelity if certain rules are followed. One rule is to limit the high frequency end of the

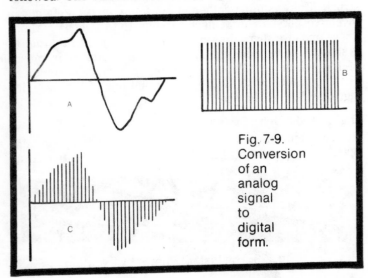

Fig. 7-9. Conversion of an analog signal to digital form.

spectrum by a low-pass filter. The sampling rate is then made several times the highest frequency component in the original signal. For example, if our signal is limited to 20 kHz by the low-pass filter, the sampling rate should be 40 to 60 thousand per second to reconstruct the original. No information is lost because the band is limited and the signal cannot change appreciably between sampling intervals.

The gradual motions we see pictured on a movie screen are made from static "samples" in much the same way. If the samples (frames) are flashed at a high enough frequency, we cannot detect the "snapshot" pulses but see only the fluid dynamic action of a live sequence.

The rudiments of the digital delay line are outlined in Fig. 7-10. The familiar analog audio signal must first be converted to digital form in the analog-to-digital converter after traversing the low-pass filter. The digitized signal then goes to a series of shift registers which do the actual delaying. These are most commonly metal oxide semiconductor integrated circuits of great complexity in a very small space. As each intrinsic element yields a delay of only 2 or 3 μsec, many elements are required to get delays of hundreds of milliseconds. Variable delay is achieved by tapping off between shift registers. Figure 7-10 is greatly oversimplified in

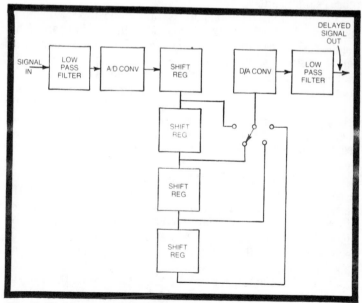

Fig. 7-10. The basic operational principles of the digital delay line.

this regard. The coarse taps shown are supplemented by other smaller banks of shift registers which are tapped for fine control. These fine control modules are duplicated so that several delayed outputs are provided.

The uses of the digital delay system in audio processing is limited only by the resourcefulness of the operator. Let us consider a few examples: Assume we have a good recording of a single violin solo. By combining this original with the same signal routed through a digital delay device set for about 10 msec delay, it sounds like two violins! After all, the 10 msec delay corresponds to the sound from a second violinist about 10 ft farther away. Routing the original signal through a digital delay line having multiple outlets, each set for the proper delay, and feeding each outlet into a console with some panpots, an entire string section can be built up from the original recording of the single violin.

The digital delay system can improve the reverberation generated by an artificial reverberator. Natural reverberation has a delay resulting from the time it takes the first impulse of sound to travel to the reflecting surfaces and return. This early reflection is then followed by a decaying series of many such reflections. With a digital delay device this delay can be added to the output of an artificial reverberator, resulting in a much more realistic effect.

Comb filters have bandpass and bandstop (rejection) responses distributed evenly throughout the audible band. The unique response of such filters has made them very valuable in certain instances. As we saw earlier in this chapter, combining a delayed signal with the same signal undelayed results in a comb filter effect. The digital delay system can be used in this way, canceling signals whose half-period is equal to the delay setting. For example, if the delay is set to 10 msec, nulls will occur at 50, 150, 250 Hz, etc., and reinforcements will occur at 100, 200, and 300 Hz, etc. These peaks and nulls can be shifted by reversing the phase between input and output. A practical use of this comb filter is in the battle against power-line hum. A recording marred by 60 Hz hum as well as 120 and 180 Hz harmonics may be salvaged by setting the delay at 16.66 msec on the digital delay line and feeding the output back out of phase. Nulls will fall at 60, 120, 180 Hz, etc. The effect on the recorded signal will be minimum because of the absence of coherent components in usual program material.

The digital delay line can be used to achieve the pseudo stereo effects described earlier in the chapter (in the phase shifter discussion). Phasing or flanging effects can also be obtained. By combining the several outputs of the digital delay

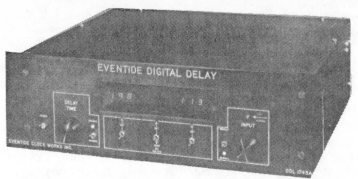

Fig. 7-11. Digital delay line manufactured by Eventide.

device in certain ways, exotic types of filters can be made up. By connecting output back into the input, a signal can be made to circulate indefinitely, which may assist in the examination of transients. The digital delay line (often referred to as the DDL) seems to be here to stay and its applications are sure to increase in number and value.

A commercially available digital audio delay system is the Eventide unit of Fig. 7-11. The standard unit offers two independent outputs, each providing up to 199 msec of delay in 1 msec steps. By means of a front panel doubling switch, up to 398 msec of delay is obtained in 2 msec steps. There are two solid-state digital readouts of delay—one for each output. Switching is quiet so that delay can be switched in and out during low-level program passages.

Another digital delay, the Gotham Delta-T 101 digital audio delay system, is shown in Fig. 7-12. It is available in its

Fig. 7-12. Gotham Delta-T Model 101 digital audio delay system.

minimum configuration as a single-channel device having delay selectable in 5 msec steps up to 40 msec. As many as five such units may be accommodated in the frame, plus seven additional delay cards of 40 msec. A precision crystal oscillator eliminates flutter and a spare plug-in card takes care of maintenance.

MUSIC SYNTHESIZERS

The output of electronic music synthesizers can be classified either as *new-sound* music or sound effects, depending on one's point of view and the expectations of the one at the keyboard and controls. Surely, these remarkably flexible instruments are capable of producing an almost limitless array of sounds and signals, justifying at least a cursory treatment under this heading of *special effects*. Our emphasis will be on methods and techniques used to show their potential for generating special sound effects.

Sound Sources

Function generators produce a variety of basic waveforms such as sine waves, square waves, triangular waves, and rectangular pulses. Each waveform contains a fundamental and, except for the pure sine wave, a train of harmonics. Another signal used is random noise with its continuous spectrum. These various waveforms are compared in Fig. 7-13. All the energy of the sine wave is concentrated at one frequency. Any distortion of the sine wave will create harmonics, but in its pure form it contains none. The square wave is rich in odd-numbered harmonics. A 1 kHz square wave would thus have harmonics at 3, 5, 7, etc., kHz. The triangular wave is rich in both odd and even harmonics, as well as the rectangular impulses. Any of these waveforms can be built up, or synthesized, by combining its characteristic harmonics in just the right magnitudes and in the right time relationships. Random noise is not a periodic wave and hence has no fixed harmonic structure. Put another way, its harmonics are continually shifting, resulting in a continuous spectrum. These waveforms provide the varied distribution of energy throughout the audible spectrum which can be shaped into an extremely wide array of sounds, but they are continuous in that once turned on, they continue their monotonous sound. We must now consider the way the monotonous can be made interesting.

Gating

By opening and closing a switch, we can produce bursts of any of these sounds. This is more conveniently achieved with electronic gating circuits which can produce a continuing series of rhythmic bursts, adjustable in both interval and duration.

Amplitude Control

Amplitude control can be achieved by turning a volume control knob. This is useful, but electronic circuits can vary the envelope of the basic signal or pulse automatically to give, for example, a piano-like transient with fast attack and slower release, plucked string sounds, and many other effects.

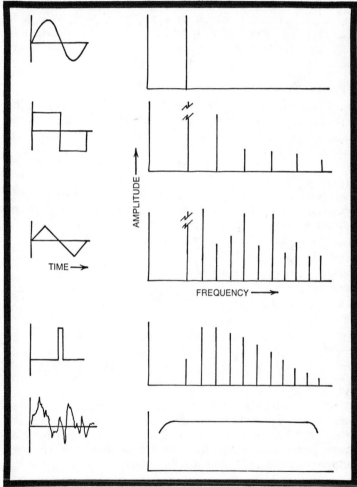

Fig. 7-13. Comparison of spectra of sine, square, and triangular waveforms, impulses, and random noise.

Modulation Techniques

Modulation is the process by which one tone or group of tones is acted upon by a second tone or group of tones. An example of amplitude modulation is *tremolo*, whereby a basic tone is made to vary at about a 6 Hz rate. *Flutter tonguing* can be accomplished electronically as well as by a musician's tongue. Frequency modulation, illustrated by the vibrato a violinist creates by rapidly moving his finger back and forth on a string, can readily be achieved electronically. Balanced modulators permit one of the input signals to be eliminated while its effect on the second signal remains. In this sense, a balanced modulator serves as a gating as well as a tone modifier. The balanced ring modulator goes even further in that both input tones are eliminated and only the newly generated modulation products go through.

Filtering Effects

Filtering can change the tonal quality (timbre) of all the waveforms of Fig. 7-13 but the sine wave. This is because the filter affects the harmonic content and the pure sine wave has none. There are fixed filters and filters which are variable as to rolloff point, slope, and maximum attenuation. There are low-pass, high-pass, bandpass, comb, and notch filters. The formant filter is a fixed filter which has a special usefulness in electronic music, making a sawtooth wave sound like anything from a violin to a bassoon. The formant filter is related to the natural acoustical filters resulting from the very structure of each musical instrument.

OTHER SPECIAL EFFECTS

Although the synthesis techniques and processes are related to the field of electronic music, they are tools for building special effects as well. For example, electronically gating white noise and passing it through a low-pass filter can produce a very acceptable gunshot sound which can be made variable from staccato to reverberant with other circuits.[6-9]

Keying Effects

We considered the Kepex as a gating device in Chapter 4 in which the gain is controlled by the level of input signal. An exterior "key" input is provided so that the gain may be controlled with a second independent audio signal. This can add novelty and interest to program material.

Variable Speed Effects

The speed of a recorder driven by a synchronous motor can be varied by adjustment of an oscillator. Playing back a

Fig. 7-14. The Eventide Omnipressor.

signal recorded at one speed at a somewhat different speed can change the pitch of voices, music, and other sounds. This can result in comical effects or, applied in carefully controlled doses, an offkey selection can be corrected. The speed of recorders having servo drive systems can also be varied by overriding the standard signal with a manually controlled signal.

Reversal Effects

The Eventide *Omnipressor*, shown in Fig. 7-14, is such a versatile device that it is being considered under special effects. It is a normal compressor, expander, noise gate, and limiter and, as such, can be used in signal processing in recording and mixdown. In addition to these functions, it can generate new effects such as infinite compression and dynamic reversal.

In the graph of Fig. 7-15 the horizontal axis represents the input level of the Omnipressor in decibels referenced to 1 mW (dBm), and the vertical axis represents output level. We establish our basic operating point where zero level input gives zero level output. If the significant controls are centered, an increase in input level of 10 dB results in an increase in output level of 10 dB and we find we have a linear input/output relationship. If compression is introduced, this operating line rotates clockwise about our operating point. As the operating line approaches the horizontal, we have a condition in which a wide variation of input level scarcely changes the output at all. This approach to infinite compression can be used as a special effect.

In Fig. 7-16 we see a linear relationship between input and output but the signs are reversed. A decrease in signal input results in an increase in output signal. This is called dynamic reversal. When applied to music, the attack/decay envelope of plucked strings and similar instruments is reversed.   Applied to speech, the effect is that of talking backwards. The creative possibilities in special effects are thus further widened.

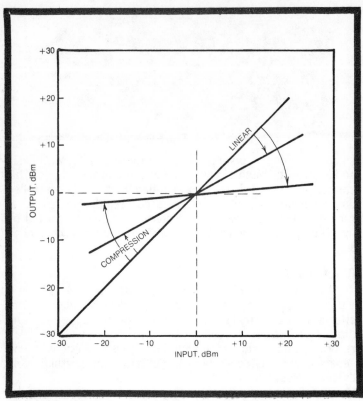

Fig. 7-15. Compression is increased as the operating line rotates clockwise around the operating point. As infinite compression is approached, a condition useful for producing special effects results. (Eventide Clock Works, Inc.)

Doppler Effects

The Leslie organ loudspeaker is usually used with the Hammond B3 organ. The tremolo-like effect is accomplished with a rotational loudspeaker mechanism which introduces a doppler effect, the pitch increasing slightly as the loudspeaker comes toward the listener and decreases slightly as it moves away like the familiar change in pitch of a locomotive whistle or automobile horn as it passes by. The Leslie doppler characteristic has been applied to vocal and instrumental program material, introducing a fascinating effect that seems to go beyond normal vibrato and tremolo. Best results are obtained by placing the Leslie loudspeaker and microphone pickup in a separate room and operating remotely. By placing it in an echo chamber, extra reverberation is available.[10]

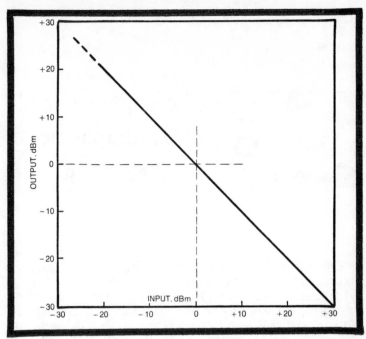

Fig. 7-16. The reversed relation between input and output level, which the Eventide Omnipressor can produce, can be used for dynamic reversal.

REFERENCES

1. Burstein, Herman. *Phasing in Tape Recording.* Audio, Vol. 57, No. 12, December 1973 (pp 54—56).

Bartlett, Bruce. *A Scientific Explanation of Phasing (Flanging).* Jour. Audio Engr. Soc., Vol. 18, No. 6, December 1970 (pp 674, 675).

3. Meyers, William. *Countryman Associates Phase Shifters How To Use Them.* Countryman Associates, Palo Alto, California.

4. Blesser, Barry and Francis Lee. *An Audio Delay System Using Digital Technology.* Jour. Audio Engr. Soc., Vol. 19, No. 5, May 1971 (pp 393—397).

5. Factor, Richard and Stephen Katz. *The Digital Audio Delay Line.* **db.** The Sound Engineering Magazine, Vol. 6, No. 5, May 1972 (pp 18, 21—23).

6. Ehle, Robert. *The Technique of Electronic Music.* **db.** The Sound Engineering Magazine, Vol. 5, No. 6, June 1971 (pp 32—33); and Vol. 5, No. 7, July 1971 (pp 32—33).

7. Bode, Harald and Robert Moog. *A High-Accuracy Frequency Shifter For Professional Audio Applications.* Jour. Audio Engr. Soc., Vol. 20, No. 6, August 1972 (pp 453—458).

8. Colin, Dennis. *Electrical Design and Musical Applications of an Unconditionally Stable Combination Voltage Controlled Filter Resonator.* Jour. Audio Engr. Soc., Vol. 19, No. 11, December 1971 (pp 923—927).

9. Nisbett, Alec. *The Technique of the Sound Studio.* Hastings House, New York, 1970.

10. Foster, Don. *Remoting The Leslie Organ Speaker.* Recording Engineer/Producer, Vol. 3, No. 4, September/October 1972 (pp 35, 37, 39).

8

Stereophonic and Quadraphonic Recording

"Signals to the left of us, signals to the right of us, volleyed and thundered" (to paraphrase an old classic) seems to characterize today's audio scene. Battles are raging on the marketing front, supported by frantic activity in development laboratories on at least three continents. Once more we see the reluctance of large companies to agree on standards, choosing rather to speculate on the hope that *their* system will prevail and that all the golden plums will land in *their* baskets.

This isn't the first time such confrontations have taken place in the audio field: we have almost forgotten such battles of the past between the 33 rpm long-play and the 45 rpm disc (both exist harmoniously side-by-side today), the vertical/lateral approach for mono, and the conflict between the two stereo disc systems—the vertical/lateral and the 45°/45° groove, the latter the victor and the standard today. Who is to say that looking to the marketplace to settle such questions is bad? After all, the market is made up of consumers who supply the financial glue that holds the entire audio structure together.

PSYCHOACOUSTIC PHENOMENA

This is a book on recording and it is not our purpose to delve deeply into the QS, SQ, CD-4, and other systems, but rather to understand something of the recording techniques upon which all of them must rely for source material.

Fusion of Sound Images

Defining the quality of a monophonic system boils down to old familiar things such as system frequency response, distortion, and dynamic range. We are inclined to take excellent performance in these categories pretty much for granted with today's advanced consumer equipment. The coming of stereophonic sound in the 1950s and quadraphonic sound in the 1970s later required the addition of a fourth quality factor—an appraisal of directional effects.

Is a true criterion of quality in recording and reproduction the faithful reproduction of some acoustic event? It once was, but the multitrack techniques so widely used today result in products which bear little, if any, similarity to any acoustic event. So much creative processing takes place in the control and mixdown rooms that only bits and snatches of the final product bear any resemblance to what happened in the studio during the recording session. There is no performance, strictly speaking, until the final mixdown is completed. Perhaps some new quality criteria are needed to cover the recording field today. We must also remember that 95% of U.S. record sales are in the pop category making use of multitrack techniques.

In two- and four-channel stereo the directional aspect of a sound is a very important part of the overall effect. These localization phenomena depend upon the psychoacoustical functioning of the ear—brain combination, and it is well for us to look at the results of several experiments scientists have performed. These experiments have to do with the fusion of signals from separate sources, broadening of the fused image, and displacement of the image.[1]

The Effect of Level

A mono source feeds a signal to two loudspeakers, LS-1 and LS-2, as shown in Fig. 8-1. The listener is positioned symmetrically as shown. If both loudspeakers radiate the same signal, in the same phase, and with equal levels, the listener will hear a single phantom image midway between the two loudspeakers. If the signal to LS-1 is attenuated, the phantom image shifts toward LS-2. Conversely, if the signal to LS-2 is attenuated, the signal shifts toward LS-1. This is the basis of the stereo panpot in placing a mono image in the desired position between two stereo channels.

Effect of Arrival Time

Consider the same arrangement as Fig. 8-1 except that delay devices replace the attenuators. Again, equal in-phase signals radiated from LS-1 and LS-2 create a fused phantom image on the line of symmetry directly in front of the listener. If the signal from LS-1 is delayed with respect to that from LS-2 at the ears of the listener, the location of the phantom

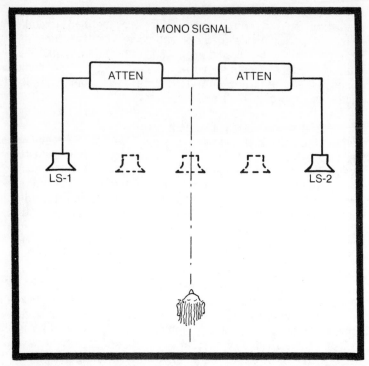

Fig. 8-1. Localization of phantom image by attenuation. If attenuators are replaced with delay lines, the same effect results.

image will shift toward LS-2 an amount depending on the amount of the delay. One easy way to get the delay experimentally is to move LS-2. Although it is usually more convenient and less expensive to use attenuation to shift the phantom image (panpot method), it can be done equally well by adjustable delay networks. It turns out that a delay of 1 msec has the same effect as a change in level of 5 dB.

Effect of Frequency Response

In Fig. 8-1 a requirement for image fusion to take place is that signals from LS-1 and LS-2 must be similar in quality. If this quality is purposely altered, the image may be broadened considerably. Let us roll off the signal fed to LS-1 above 1 kHz and roll off the signal to LS-2 below 1 kHz. The listener of Fig. 8-2, located symmetrically between the two loudspeakers, now has the illusion that both LS-1 and LS-2 are delivering full power for a wide band. Actually, the listener does not have to stay in a limited spot to get this effect, especially if the loudspeakers are far enough away so that his moving about does not result in appreciable changes in relative levels from the two loudspeakers. Sharper skirts on the dividing network do not destroy the effect, but they result in LS-1 sounding wideband with low-frequency emphasis, and LS-2 sounding wideband with high-frequency emphasis. Thus, we see that a difference in the quality of the signals from the two loudspeakers tend to broaden the phantom image and give a pseudo stereo effect.

As stereo records first appeared on the market, many hi-fi enthusiasts sought ways to extract a pseudo stereo effect from their mono records. One simple expedient was to feed the same mono signal to the inputs of two amplifiers, each driving a separate loudspeaker. By turning up the bass and turning down the treble tone control on the left channel and doing the reverse on the right channel, the situation of Fig. 8-2 can be approximated and a semblance of stereo effect achieved.

An elaboration of this method, exploited commercially (Orban Parasound and Countryman Associates, etc.), is to

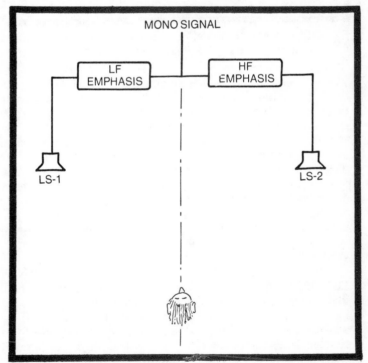

Fig. 8-2. Pseudo stereo effects by adjustment of quality of signal.

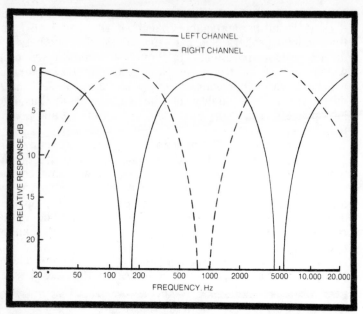

Fig. 8-3. Pseudo stereo effect through use of comb filter, feeding some bands to left loudspeaker, others to right.

utilize several bands, feeding some to the left and the others to the right channel. Figure 8-3 illustrates a five-band approach, three to the left and two to the right channel[2]. These bands are generated by complementary comb filter techniques (see Fig. 7-2). By adjusting the bands, the sounds of some musical instruments can be positioned left or right because of the differences in their spectra, heightening the directional effect. It is also possible to divide the nondirectional low-frequency energy between the two channels to make more effective use of the two amplifiers. Such devices have been used professionally in reprocessing old mono records for new pseudo stereo release, but this is generally no longer profitable.

In Fig. 8-4 a 180° phase shift is applied to the signal driving LS-1. This is accomplished simply by reversing the leads to that loudspeaker. The phantom image perceived by the listener is now inside or near the back of his head. If, in addition to shifting the LS-1 phase by 180°, the signal of this channel is also attenuated, the phantom image position now shifts to a point beyond LS-2, as shown.

We have examined several of the various factors affecting the fusion of sound from two loudspeakers fed from a

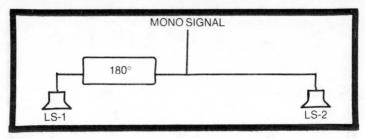

Fig. 8-4. Phantom images can be localized to positions other than between the two loudspeakers by use of phase shift.

monophonic source. What has this to do with stereophonic and quadraphonic recording and signal processing? Well, for one thing, the various quadraphonic matrix systems rely on attenuation and phase shifting for their effect. Further, the master tape is built up from many mono sources in the usual multitrack approach, often combined with some stereophonic signals. These mono sources are positioned at will by panpots.

At this point we should broaden our concept of the common array of four loudspeakers as shown in Fig. 8-5 and begin to

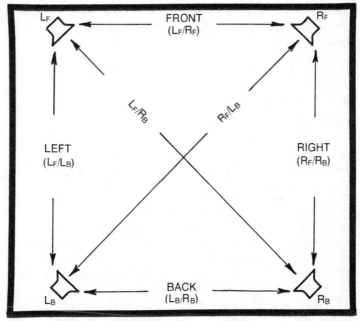

Fig. 8-5. Four loudspeakers in a quadraphonic system as six stereo pairs.

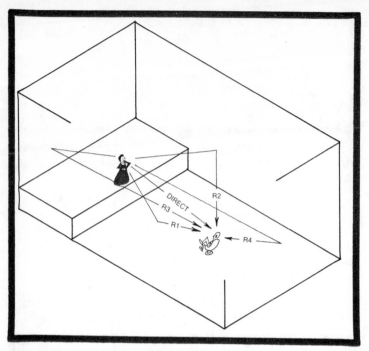

Fig. 8-6. Reflected components comprising the "early sound."

think of it as six stereo pairs as indicated. If we are considering channel separation, for example, here are the six different components of the problem.

AMBIENCE, WHAT IS IT?

Let us consider the room arrangement of Fig. 8-6 with an enthusiastic sound source on the stage and a lone listener in the audience area. We need a short-duration impulse in our analysis, so we direct the sound source to let out with a staccato blast. If, as the man said, "a short line is the straightest distance between two points," the first arrival of the sound pulse at the listener's ears will be by way of the path labeled *direct*. A bounce off the floor follows closely in time because of the small difference between the direct and R_1 path lengths. Reflection R_2 will be correspondingly later and R_3 and R_4 still later. We have considered only the first few reflections, but it is obvious that these will be followed by a torrent of other reflections of generally lower amplitude because of the greater distance traveled and losses due to multiple reflections.

Let us assume that the above room is of such size and construction as to yield the time and amplitude configuration shown in Fig. 8-7. Reflections R_1, R_2, R_3, and R_4 are lower in amplitude than the direct ray because they have traveled farther and have been reduced by reflection losses from floor and walls. We have taken as zero time the instant the direct ray reaches the listener. In other words, the graph of Fig. 8-7 has been drawn strictly from the point of view of the listener. This is important because we are considering certain psychoacoustic effects which drastically affect the way the listener hears these various reflected components.

Helmut Haas has generally been credited with the discovery of the *Haas effect* or the precedence effect in our audio literature. We are indebted to Haas for some excellent research carried out for his doctoral dissertation at the University of Gotingen under the eminent E. Meyer, published in 1951[3]. However, Haas wasn't first, for the effect had been studied by a string of well known scientists as far back as Joseph Henry in 1849, as pointed out by Gardner[4]. The Haas effect or precedence effect is the name given to the phenomenon of the human hearing mechanism integrating all the sound arriving within a certain time period. In our example, any reflected energy arriving within about 35−40 msec (after the direct sound) is perceived as coming directly from the source. In Fig. 8-7 we note that reflections R_1, R_2, R_3 and R_4 all arrive at the ears of the listener within 30 msec of the direct sound. The ear−brain combination perceives all of these as direct sound coming from the direction of the source even though some of it comes from the sides and from the back. The reflected components R_1 through R_4 are heard not individually but only as a sort of support and enhancement of the direct sound.

Sound arriving later is perceived as echoes or, in our case, as reverberation arriving from every direction. Within 30 msec the ear definitely gathers all information together and identifies it as direct sound. After 70 or 80 msec the integrating effect is not active. In the transition zone between there is some integrating effect, but it is diluted. This explains why some writers refer to the Haas effect acting up to 30 msec, some to 40 msec, and some even take 50 msec as the upper limit.

Beyond the transition zone the rays from the sound source which have been reflected from several surfaces are perceived for what they are: random arrivals of random amplitudes spread out in time. With only modest oversimplification, we can then divide all the complex sound

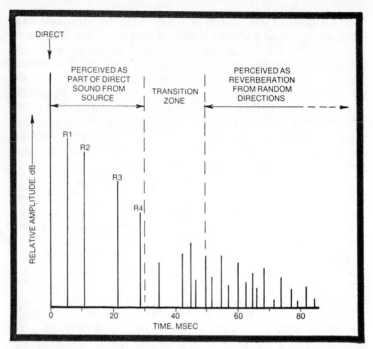

Fig. 8-7. Reflected components arriving within the first 30 msec are heard as part of the direct sound, later components as spatially distributed reverberation.

field in the room set up by the single staccato impulse into two parts, the *early* sound and the *later* sound. The direct sound is made louder by the integrated reflections and it all appears to come along the direct sound path. The later sound is spatially diffused reverberation which tails off as determined by the acoustics of the enclosure.

The word ambience has the basic meaning of surrounding, circumfused on all sides. The ambience information in sound recording and reproduction must then be associated with the later sound which is characteristic of the space in which a recording is made or an acoustical event takes place. It gives us the impression of its size. We cannot overemphasize the importance of ambience in recording. We must give top priority to making our artificial reverberation give the impression of natural ambience.

CLASSICAL STEREOPHONIC RECORDING

Good stereophonic recording requires an adequate microphone system. An adequate microphone system involves careful placement of well matched, high quality microphones. The placement of the microphones is very much a function of the size of the musical group, the repertoire, and the acoustical characteristics of the hall. Two microphones are employed to convey directional information and a sense of spaciousness. These microphones may be arranged in a *time /intensity* system spaced some distance apart, or in an *intensity-difference* system with the two microphones coincident.

Spaced Microphones

Two microphones spaced some distance apart enable the conveying of spatial information because of the difference in both the intensity of sound and the difference in the time of arrival of the sound falling on the two microphones. The basic ground rules for the placement of the microphones go right back to the precedence effect. As sound travels approximately 1.1 ft/msec, separating microphones 55 ft results in a delay of about 50 msec between the outputs of the microphones, creating an undesirable echo effect. A microphone spacing of less than 33 ft (delay less than 30 msec) would place the two microphone signals within the integrating time of the human ear which, when reproduced, would give an accurate positioning of sound sources.

The listening angle approach relates microphone placement for the recording process to the geometry of the home reproduction setup[5]. The home hi-fi setup is often dictated more by existing living room conditions than by any theoretical considerations, but the 40° arrangement of Fig. 8-8 may be reasonably representative and helpful as a general guide. Let us now take this 40° angle into the concert hall or studio where the recording is to take place and hope that the "ideal" listening position and the 40° angle will embrace the entire musical group. With this rationale, the left- and right-channel microphones should be on the 40° lines; the center-channel mike, if one is used, midway between them. The line of microphones may now be moved closer to or farther from the musical group until the proper balance of direct and reverberant sound is obtained. If the microphones are too far from the source, the sense of directionality is lost and clarity of signal suffers. If too close, the ambience of the hall is lost. Experimentation is the only way to determine the proper distance for the microphones.

Directivity patterns of the microphones are, of course, very important in establishing the ratio of direct to reverberant sound. Omnidirectional microphones would be

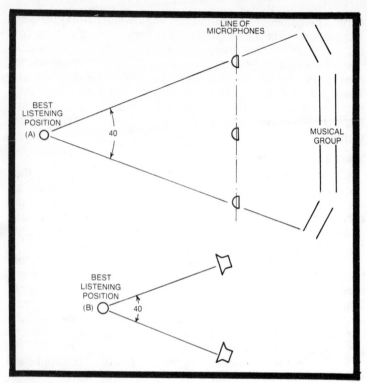

Fig. 8-8. The listening angle approach to stereophony: (A) recording and (B) reproduction.

most useful with wide musical groups and in halls having very short reverberation times. If there is a problem with reflections from side walls, figure-8 patterns with the nulls aimed at the sides might be suitable. Cardioid microphones are very commonly used in practical situations in which the hall is more reverberant than one would choose. Aimed forward, cardioids would discriminate against reverberation from the rear of the hall, tending to keep this troublesome factor in balance.

The spaced microphone technique may be employed with smaller musical groups as well, such as for chamber music in smaller rooms. The geometric guides given may be scaled up or down to fit the job at hand. If a center mike is used the center track signal may be split and applied equally to the left and right channels.

Many recording engineers have little faith in such an arbitrary listening angle approach. Instead they would rely more on a knowledge of the Haas effect and their own experience. They know that when the microphone spacing is great, there will be a great difference in both intensity and time of arrival of sounds falling on the microphones from different parts of the musical group. When the spacing of the microphones is very small, the *coincident* microphone arrangement results where time-of-arrival differences are small and microphone response depends on the directional pattern of the microphone. Experienced recording engineers realize that greater spacing of microphones tends toward an open, spacious effect, and that closer spacing tends toward a tight, constricted effect. Too great a spacing also tends toward a hole in the middle (which can be minimized by the use of a center microphone). A spacing of 3—6 ft has been found satisfactory for a large orchestra.

Coincident Microphones

In 1931 Blumlein obtained a British patent on the coincident microphone principle, hence the *Blumlein pair* designation which appears now and then in audio literature. In the U.S. the use of the coincident microphone principle is commonly called the MS system. Two condenser mikes are typically mounted on the same axis, one above the other in the same housing. One of the microphones, of the cardioid type, is the *mid* microphone, which covers the musical group. The other has a figure-8 pattern oriented parallel to the musical group (the *side* microphone) as shown in Fig. 8-9. A variation

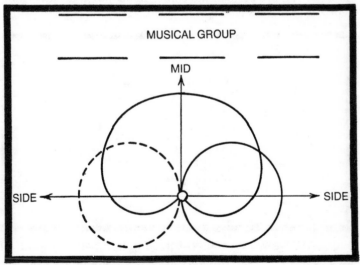

Fig. 8-9. The MS (mid side) microphone arrangement for stereophonic recording.

of the MS microphone replaces the cardioid element with an omnidirectional element.

The MS coincident microphones are mounted centrally in front of the musical group. The cardioid mid microphone covers the entire group with its wide forward lobe (left + right + center or L + R + C). A characteristic of the figure-8 microphone is that signals picked up on one lobe are 180° out of phase with those picked up on the other lobe (no center, but left minus right or L − R). The outputs of these two coincident microphones are combined in a simple network composed of two carefully matched and balanced microphone transformers connected as shown in Fig. 8-10. The sum components for the left channel are obtained by connecting secondaries S_1 and S_3 in series aiding, and the difference components for the right channel by connecting S_2 and S_4 secondaries in series opposition. In this way the mid microphone output is added to the side microphone output

$$(L + R + C) + (L - R) = 2L + C$$

to obtain the left channel with reduced center information. Similarly, the difference between the mid and side outputs

$$(L + R + C) - (L - R) = 2R + C$$

yields the right channel with reduced center-channel information—just what is needed. It is possible to inject some control of channel separation at this combining stage if it is desired. The presence of the center information assures good reproduction of a stereo disc in mono.

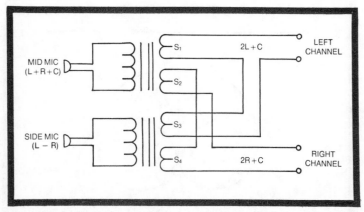

Fig. 8-10. Network (matrix) for combining the outputs of two coincident microphones in the MS system to obtain the sum components for the left channel and the difference components for the right channel.

X-Y Microphones

Another system of using coincident microphones in stereo recording is called the X-Y system. In this system one directional element of the coincident pair picks up sound predominantly from the left and the other from the right. In Fig. 8-11A two microphone elements having figure-8 patterns are mounted in the same housing very close together, and with

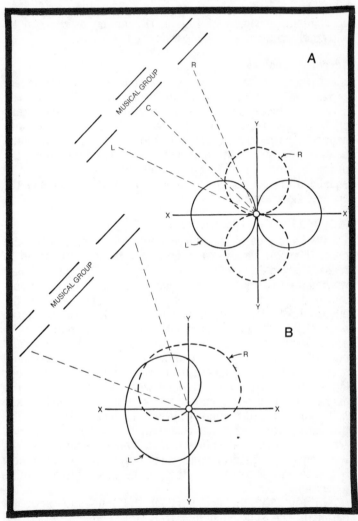

Fig. 8-11. Using two coincident microphones with (A) figure-8 patterns in the X-Y system, and (B) cardioid pattern in the X-Y system.

171

the axes oriented 90° to each other. Sound from the left side of the musical group will give a high output in mike L and a low output in mike R. Sound from the right side of the musical group will give a low output in mike L but a high output in mike R. A sound from the center results in equal outputs of the L and R mikes. Out-of-phase components which exist can be manipulated in a network similar to that in Fig. 8-10 to make the final stereo image broader or narrower. The coincident cardioid microphones of Fig. 8-11B can be used when an extremely wide working angle is desired.

Accent Microphones

An alert recording engineer soon senses the need to accent certain sections of a musical group or orchestra. The percussion, brass, and bass string sections located in the rear often can benefit from the placement of accent microphones near them. Mixing accent signals into the main program must be done with great skill and restraint to avoid damaging the primary stereo effect. Sometimes the accent signal is equalized to give a high-frequency edge to the sounds of the accented instruments.

QUADRAPHONIC RECORDING

We have treated the stereophonic recording of classical music in some detail. What new factors are introduced to record the same type of music quadraphonically? Instinctively we realize two more microphones are necessary to supply information for the two additional loudspeakers of the quadraphonic reproducing sustem. If we look to the stereo microphones to supply information of a direct nature as to source location, musical texture, etc. of the orchestra spread out before the listener, the other two microphones should supply ambience information for the side or rear loudspeakers. This must be done with care to optimize the overall effect on reproduction, for both pairs actually handle both types of information. Unless it is desired to place the listener in the midst of the orchestra, what we have discussed concerning stereo mike placement still holds generally for the front mike in quadraphonic recording. The question now becomes one of sorting out the factors affecting rear mike placement.

In Fig. 8-12 the geometry of the four-mike array is detailed. The spacing of the front microphones is S_F, the spacing of the back microphones is S_B, and the distance between front and rear mike arrays is D. The front mikes are positioned according to the principles outlined previously for stereo recording, except that now they can be adjusted for optimum pickup of the orchestra but still sharing the burden of conveying ambience with the rear mikes. The Haas effect warns us that a separation D of 50 ft would certainly create disturbing echoes in the rear loudspeakers. For this reason, D should be no more than about 20 ft. If distance D is held within the Haas limit, the rear channels in reproduction can operate at levels approaching those of the front channels without disturbing localization effects. This makes for a broader listening area as the record is reproduced.

Microphones having a cardioid pattern are suitable for the back pair. If the hall is very live, they may be turned more toward the orchestra; if very dead, more toward the back. The rear mikes pick up both ambience and direct sound from the musical group, and the apportioning of these by adjustment of the distance D must be determined by experiment.

One of the main reasons for the rear microphones is to reproduce the randomness of the reverberation, and this

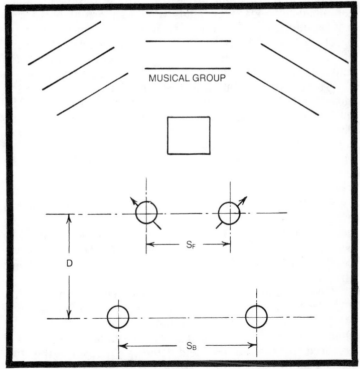

Fig. 8-12. Microphone arrangement for quadraphonic pickup of large musical group.

affects the spacing S_B. The closer they are together, the less the randomness (the greater the coherence) between their signals.

Eargle has given microphone spacings which have been used in recording a large orchestra covering an area 50 ft wide and 30 ft deep[6]. For spacing, they are:

Spacing:
S_F—3−6 ft
S_R—8−6 ft
D—20 ft

And for height:
Front pair— 12−15 ft
Back pair— 15−20 ft

Close Microphone Technique

The traditional approach to recording classical music has been to consider ambience of the hall an important part of the recording. With popular music the acoustic environment plays a less important role and close-mike techniques have therefore been applied widely in that area. Surely, to approach the recording of a symphony orchestra with the same techniques commonly applied to rock bands would be a formidable task. It will be interesting to see if close-mike techniques ever fully penetrate the bastions of classical music recording. On the other hand, close techniques are being used extensively for music such as that of Boulez, Stockhausen, and Birtwhistle.

The entire orientation of this book has been toward the close-mike, maximum separation, multitrack approach. The following chapter is given over completely to a detailed description of actual popular music recording and mixdown sessions. For this reason no further consideration of close-mike techniques is appropriate at this point.

Panpots

In Fig. 8-1 we saw how it was possible to shift the phantom image across the field by adjusting the relative levels of the two channels. In practice, the panpot is the device used to position monophonic sources in the stereo field. The same can be done by adjusting delay devices, but attenuators are cheaper than delay lines.

The common stereo panpot arrangement is shown in Fig. 8-13. The two attenuators of Fig. 8-1 are combined into a single attenuator R_p between the left and the right channels. If the sliding contact of R_p is at dead center, the signal of channel 1 is centered in the stereo field, midway between the two loudspeakers. If the sliding contact is moved to the extreme left, channel 1 sound is heard on the left loudspeaker, and so on. By moving this slider, channel 1 may be positioned at will.

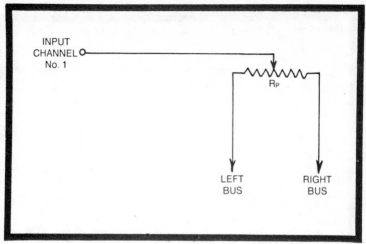

Fig. 8-13. The common stereo panpot arrangement.

In Fig. 8-14 a panpot bank, accommodating eight input channels, feeds two stereo lines. Any of the input signals may be freely located anywhere in the stereo perspective with this simple arrangement. The panpot for each input channel is normally located on the input module panel of the console.

Stereo is basically a one-dimensional system as sound sources are made to appear at various positions along a single line, although the illusion of depth can be partially achieved by adjusting levels, adding reverberation, etc. For quadraphony, the director has in mind a two-dimensional acoustic perspective field involving left and right as well as forward and back dimensions. In his mind he places each performer somewhere in the acoustic perspective field and it remains only to achieve the effect he has envisioned. This is done by routing each performer's channel to one of four outputs, or someplace between them. Bus 1 may represent left front, bus 2 right front, bus 3 left back, and bus 4 right back. With the quadraphonic panpot (quadpot) the director can move the sounds of performers around at will, assuming he has adequate track separation to work with.

The quadpot is somewhat more complex than the stereo panpot, yet it operates on the same basic principle of Fig. 8-1. It is well to have an understanding of its operation. Figure 8-15 is a schematic representation of the way the four potentiometers of each panpot are connected electrically. All that remains is to devise a joystick to move these potentiometers so that the proper psychoacoustic result is achieved.

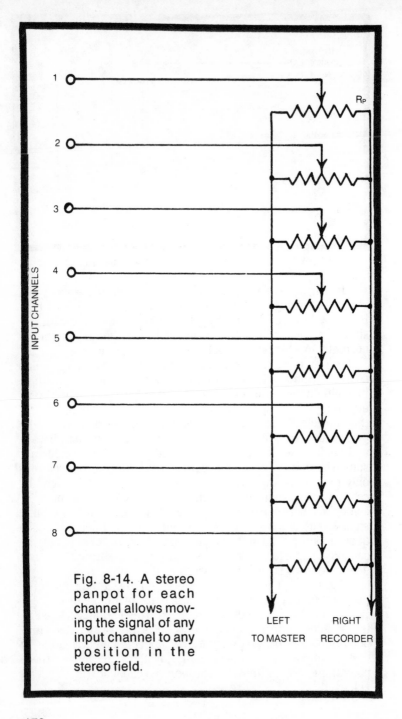

Fig. 8-14. A stereo panpot for each channel allows moving the signal of any input channel to any position in the stereo field.

INPUT CHANNELS

1

2

3

4

5

6

7

8

R_P

LEFT

RIGHT

TO MASTER

RECORDER

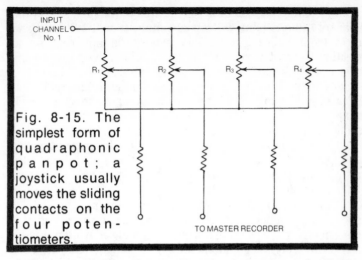

Fig. 8-15. The simplest form of quadraphonic panpot; a joystick usually moves the sliding contacts on the four potentiometers.

TO MASTER RECORDER

Figure 8-16 shows the way *Spectra-Sonics* has solved this problem in its Model 904P pan control. The joystick moves a metal sphere to which the sliding elements of the four linear potentiometers are mechanically affixed. If the joystick is on the straight-up neutral position, the sliding contacts of all four potentiometers are in the center position and the sound of that channel would apparently be in the center of the room. If the joystick is pushed straight ahead to its extreme position, the

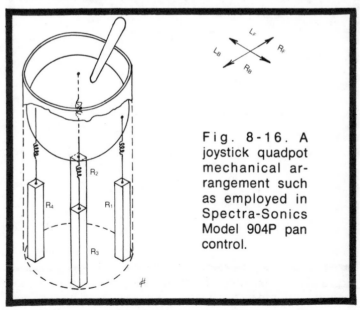

Fig. 8-16. A joystick quadpot mechanical arrangement such as employed in Spectra-Sonics Model 904P pan control.

sliding contacts R_1 and R_2 are connected to the high side of the channel 1 line, sending full signal voltage to both left front and right front loudspeakers. This moves the apparent sound position to front center. If the joystick is moved toward the operator to its extreme position, the reverse is true: R_3 and R_4 sliders are now at the high side of the channel 1 line and R_1 and R_2 are minimum. Now the left back and the right back loudspeakers are equally driven and the apparent source of sound is directly to the rear. Moving the joystick to the left front directs maximum voltage to the left front loudspeaker and places the apparent location of the sound there. By rotating the joystick in a circular motion, the sound of channel 1 can be made to encircle the room.

The Tascam Model 107 quadraphonic panpot is also operated by a joystick, but it is somewhat more complex electrically in the effort to improve channel separation. Figure 8-17 shows eight potentiometers in a single quadpot, each panning potentiometer having a compensating potentiometer associated with it. Four such quadpots controlling four input channels are shown schematically in Fig. 8-18. Normal rotary potentiometers are used in the mechanical arrangement as shown in Fig. 8-19. There are other methods of panning the input channels into the quadraphonic configuration, but these

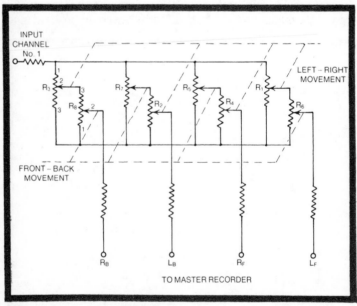

Fig. 8-17. Circuit diagram of quadpot utilizing a second potentiometer in each channel.

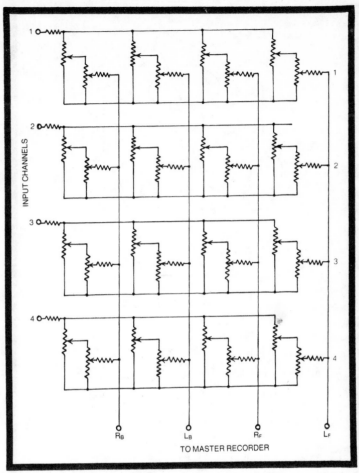

Fig. 8-18. Replication of quadpot of Fig. 8-17 to control a multiplicity of input channels.

give us sufficient background to use panpots effectively in our mixdown and other work.

MONAURAL/STEREO

Although the disc market has gone exclusively to stereo and quadraphonic releases, the mono mode still prevails in at least one medium—broadcasting. AM broadcasting as well as much of FM broadcasting is mono. This places great importance on being able to obtain good mono signals from both 2- and 4-channel stereophonic discs. Let us look at the stereo—mono compatibility problems briefly to see how they

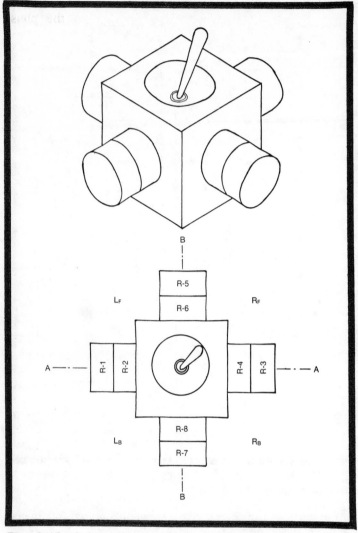

Fig. 8-19. A joystick quadpot mechanical arrangement employed in the Tascam Model 107 module.

impinge on recording techniques. Eargle gives several suggestions for improved compatibility[7]:

- Use spaced microphones with caution for solo instruments or for accent when there is appreciable overlap in their patterns. When combined in mono, selective reinforcement and cancellation occurs, degrading the mono quality.

- Use coincident microphones (such as the MS), as these rely on intensity variations and escape the phase problems of spaced microphones.
- The use of out-of-phase coherent signals in stereo to achieve unusual localization effects outside the bounds of the two reproducing loudspeakers invariably creates problems in mono.
- Use modest monitoring levels during multitrack mixdown to avoid imbalances and alternate stereo and mono monitoring in the interests of mono compatibility.
- Make sure that all stereo transmission and recording channels are balanced and aligned.
- Keep levels of low-frequency information within the bounds of trackability on inexpensive playback equipment.

QUADRAPHONIC MIXING

Techniques of 4-channel mixing will probably be as slow in emerging from stereo as stereo was in emerging from mono. The proper handling of image placement and reverberation dominate quadraphonic mixing. If the program is built up from essentially mono sources, image placement depends wholly on quadpot action; unfortunately, it is not perfect. As its action is based only on intensity, it affords precise location only at the four corners; so phantom images, particularly at the sides, are often unclear and unsteady. Another outgrowth of the quadpot problem is that good listening is realized only over a very small area.

Realistic 4-channel reverberation can be a balm for many of these problems. Cunningham and Swedien[8] suggest that if the soloist is placed in left front, then feeding this signal delayed—say, 5, 10, and 15 msec—to the other three channels at decreasing levels would provide simulated early sound to support and strengthen the soloist. Feeding the signal delayed 30 msec to a reverberation chamber with a 4-mike return would provide the incoherent reverberation which can be mixed in to closely simulate the original effect. The soloist stays solidly in position and the listener can move about without losing the intended effect. Unfortunately, most consoles do not yet provide for such needs but good artificial reverberation devices and quadpots that even position images above the horizontal plane are undoubtedly just around the corner.

EXPERIMENTAL 4-CHANNEL DRAMA

The British Broadcasting Corporation is now issuing quadraphonic recordings to overseas radio organizations in discrete or matrixed form. One such production, Shakespeare's *The Tempest*, made in quadraphony, has been described by production engineer Adrian Revill[9]. A cluster of four Neumann KM86 cardioid microphones was used with their axes at right angles. The music was recorded with the three singers fairly close to the microphones and the instruments arranged around the microphone in a circle. Each instrument was also given a close microphone and these were fed to the remaining 4 tracks of an 8-track recorder to provide flexibility in adjusting perspective, level, and position of the instruments in the mixdown. For voice recording, the characters were also arranged in a rough circle around the microphone cluster. Each character requiring it had a separate microphone for asides which was panned appropriately for position.

A disadvantage of such quadraphonic recording became obvious in that there was no "dead area" for a much-needed cough or a rumbling stomach. The studio acoustics were also greatly exaggerated. It was agreed that such recording requires a very dead studio, extremely quiet air-conditioning equipment, and great isolation from outside noise.

What spot effects were needed were achieved by the effects man standing behind the actor and rattling a chain or drawing a sword or whatever was required. Quadraphonic sound effects were synthesized from stereo effects by using long tape delays, which worked well for sustained sounds such as wind and rain. For thunder and creaking of the ship's rigging, mono sources were panned around and from side to side to achieve the desired effect. The storm scene proved to be the most difficult, but apparently one of the most successful—one critic confessed that he felt drenched after hearing the play!

One unusual effect desired was to give the impression that the voices on the ship were spinning around as the ship sank. To achieve this the mike cluster was rotated by tugging on a piece of elastic fastened to one corner of the cluster. When Caliban fell flat to hide under a gaberdine, the actor mumbled and moaned into the studio carpet with excellent effect. When King Alonso rushed toward the open sea, the studio doors were opened and he ran shouting down the corridor for another spectacular effect.

The music tracks were edited and reduced first to a 4-channel music master. The speech tracks were combined with effects and reduced to a similar speech master. These were then combined for the final 4-track master. More than merely an experiment in quadraphonic drama, it pioneered multitrack techniques in radio drama.

MIXDOWN

Mixfix has come to be the dominant philosophy of multichannel/multitrack recording. If anything is less than ideal, "we'll fix it in the mix," after the expensive musicians have left and some degree of calm has replaced the frenetic pace of the recording session. Of course, there are limits—not everything can be fixed in the mix. But to the director and the recording engineer, there is great comfort in the second chance that a mixdown affords.

The recording engineer and the director must have the mixdown in mind from the beginning. The way the instruments are arranged, the way the microphones are placed, all must be carried out according to some logical plan, and that plan is the mixdown plan. For instance, let us assume that the eight original tracks are to be mixed down to a stereo master. The recording engineer must have in mind the placement of each sound source in the final stereo field as he sets up the session. If drums are to be placed in far stereo right, there must be very good separation between the drums and the electric guitar in stereo left or he will be unable to achieve the envisioned plan. And the difference between success and failure may be something as simple as the careful direction of the guitar microphone null toward the drums. On the other hand, if the vocalist and acoustic guitar are both to be near the center of the field, less separation between their tracks is required.

The various microphone signals routed through their respective input channels of the recording console are given a minimum of processing as they are being recorded. Only those things are done to the signal in recording which improve the dynamic range, the signal-to-noise ratio, or which compensate for deficiencies in the source. If a code/decode noise reduction system is employed, certainly the initial noise coding step in mixdown is a must for lowest noise in the master transfer. It is in the mixdown that the bulk of signal processing takes place. Temporary mixes are for monitoring, for reference, or for foldback purposes only; they do not affect the original sound as it is being recorded.

The Mixdown Console

In the smaller studios, the same console is generally used for both recording and mixdown. There is no functional compromise in this as far as the console itself is concerned; temporary mixdowns, as we have noted, are an important part of the recording process for monitoring, cue, and overdubbing. The compromise comes in the scheduling. While the board is being used for recording, it cannot be used on a mixdown job. If the recording jobs are few and far between, this may be a satisfactory arrangement, but if plentiful, this may be a costly procedure. The most flexible arrangement is to have a separate room or rooms and equipment reserved for mixdown, and this requires tape playback and other equipment as well as a console. This second room could very well be the control room of a second studio.

The mixdown (or remix) console can be considerably simpler than the recording console. Signals come in at line levels from tape machines rather than at microphone levels, which eliminates the need for preamplifiers and certain microphone controls. Fewer channels may be required as only the number of tracks recorded need be handled, which is ordinarily much less than the number of microphone channels. There may be other savings in the smaller number of level indicators, lack of foldback circuits, smaller desk size, etc., but few manufacturers offer stock consoles for mixdown use exclusively, presumably because of limited demand for them. Studio operators apparently judge the sacrifice in flexibility too great to be offset by the lower cost. An intermediate sized stereo remix console offered by Audio Designs is shown in Fig. 8-20. It is only a yard wide for 16 channels, with an extra input for "the instrument that wasn't in the studio."

Tape Position Automation

Useful both in recording and mixdown, automatic tape position locator devices are finding an ever-increasing application in recording studios. These digital devices are activated by a sensor tied to the speed of the tape and are provided with memories, readout counters, and various control buttons for command input. An accurate written log of every take, overdub, and action of the recording engineer is basic for the mixdown operation. Without it, endless hours would be spent in searching and listening to find out what is on the tape, and the valuable original would be degraded in the process. The automatic tape position device does not affect the importance of the written log; in fact, data from the locator for starts and stops are entered in the log. When a given selection is required for playback, position data from the written log is entered into the tape position device which automatically runs down to the desired position.

A typical tape position locator, the 3M *Selectake* is considered an accessory for the 3M professional mastering recorders. Four-place digital readouts indicate recorder tape position in minutes and hundredths. The unit will search for

Fig. 8-20. A console especially designed for mixdown, the Audio Designs Model 1641. Simultaneous 4-, 2-, and 1-channel mixdowns are possible from 8- or 16-channel source. Features include high-level inputs, four reverb channels, four-knob equalization, and solo controls on all inputs.

and automatically locate a preselected tape position. After the desired tape position is entered into the preselector and the appropriate button (forward or reverse) is pushed on the recorder, a button is pushed on the locator and the tape runs down to the desired position and stops. The advantages of such a device in convenience, accuracy, and saving of time can be great indeed.

Mixdown Automation

A mixdown can be quite a complicated and nerve-wracking operation. Let us consider a typical 16/4-channel mixdown. There are 16 faders, 16 reverberation send controls, 16 equalizer modules (each with three or more controls), 16 uncalibrated quadpots, and to this must be added the reverberation return level controls, panners, master faders, etc. The operator must maintain compatibility with stereo and mono as he mixes the 16 tracks. Several of the

185

tracks are overdubbed and they must be handled differently in regard to special effects, deletions, going in and out of reverberation, occasional pan movements, etc. It all adds up to an overwhelming number of things to remember; and if an operator is to get through such a mix he must rehearse, rehearse, and rehearse. And then just as he is ready for what promises to be the final mix the director says, "Let's bring the drums up three dee-bee in that solo spot, put more highs on the vocal, and reduce the reverberation in the left front channel!"

The application of automation to recording is, perhaps, somewhat limited because of the artistic aspects of the operation; but its application in mixdown allows greater freedom from routine operations and hence a greater opportunity for artistic expression. Because of this, recording studio operators are showing great interest in highly sophisticated automatic control equipment.[10, 11, 12]

There are three basic categories of programmable functions in the completely automated control of a console: (1) level functions, (2) switching functions, and (3) equalization functions. At this time only equipment for the control of level functions and limited switching is available, but it is only a matter of time before completely automated switching and equalization are available.

One approach to the automated control of level is the VCA, or voltage-controlled amplifier (Fig. 8-21). The gain of this amplifier is determined by the magnitude of the DC control voltage applied to it. Instead of the fader acting directly on the audio signal, it is done indirectly—the fader setting

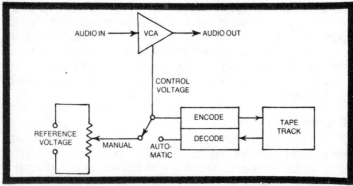

Fig. 8-21. One approach to the control of channel gain in automated mixdown is the voltage-controlled amplifier. The digitized control signal recorded on an empty track can control the channel level when played back and decoded.

determining the magnitude of the control voltage in the *manual* mode, and the decoder supplying the control voltage in the *automatic* mode. The control voltage from the fader is applied not only to the VCA, but to the encoder which digitizes it and sends it to a spare track on one of the multitrack recorders being used. In the automatic mode, this recorded control signal is processed in the decoder and used to control the VCA.

In a typical installation adapted to achieve automated gain control with an existing desk, the faders of the console are utilized to produce voltages exactly proportional to the relative fader positions. The other equipment (VCA devices, logic switching, and programer) is mounted in a rack. No basic modification of the console is required, but it now has a new capability of remembering fader settings and changes. In the usual mixing application, two spare tracks on the master tape are required for data recording. One track records the digital data corresponding to the fader settings of the 16, 24, or more channels. At each pass of the tape, changes may be recorded on the second track until the desired mix is obtained. When all the changes are in and combined with the unchanged data, a permanent memory track guarantees the same mix at every playback. The operator has full freedom to punch back and forth between stored level data and real-time fader levels as the control tracks are built up.

New integrated circuit modules are fast appearing which show promise of further audio control automation in the near future, including the digital programmable switch[13]. It would seem that simple switching circuits would present only minor problems, but this is not necessarily the case. The troublesome contact bounce of mechanical relays has its counterpart in solid-state devices; and the design of switches that are fast, smooth, free from interaction with other circuits, and economical constitutes a major engineering problem. However, the monolithic balanced modulator can be made into a digitally programmable switch having many features including freedom from switching transients. But the digitally programmable switch turns out to be more than just an exotic way to replace the double-pole-double-throw switch. It can also take a digital word such as might be generated during an automated mixdown session and directly perform an attenuation function without benefit of encoders, decoders, or voltage-controlled amplifiers. It can also serve as a programmable gain amplifier. In this one device, then, we see promise of automated switching, attenuation, and amplification functions for the console of the future. Add to this the prospects which another integrated circuit offers, the multichannel operational amplifier, and we see great hope for the future. This type of integrated circuit is also an effective link between analog signal processing and the digital control of analog audio signals.

Acoustics of the Mixdown Room

Doing the mixdown in the control room, in one sense, is an ideal arrangement in that the acoustics for the recording and for mixdown are identical. In the smaller studios this is the generally accepted way of doing things; therefore, if the control room acoustics are right, there is no problem for mixdown. In the larger studios in which the pressure of scheduling requires that mixdown be done in other rooms, care should be taken to insure that mixdown and recording acoustics are compatible so that the operators doing the two jobs hear the same thing. The critical factors in acoustical treatment in control rooms have been discussed in Chapters 6 and 12; all of these factors having to do with the listening function apply to the mixdown room as well.

REFERENCES

1. Gardner, Mark. *Some Single- and Multiple-Source Localization Effects.* Jour. Audio Engr. Soc., Vol. 21, No. 6, July/August 1973 (pp 430–437).

2. Orban, Robert. *A Rational Technique For Synthesizing Pseudo-Stereo From Monophonic Sources.* Jour. Audio Engr. Soc., Vol. 18, No. 2, April 1970 (pp 157–164).

3. Haas, H. *The Influence of a Single Echo on the Audibility of Speech.* Jour. Audio Engr. Soc., Vol. 20, No. 2, (pp 145–159).

4. Gardner, Mark. *Historical Background of the Haas and/or Precedence Effect.* Jour. Acous. Soc. Am., Vol. 43, June 1968 (pp 1243–1248).

5. Silver, Sidney. *Stereophonic Microphone Techniques.* **db.** The Sound Engineering Magazine, Vol. 3, No. 2, February 1969 (pp 21–24).

6. Eargle, John. *On The Processing of Two- and Three-Channel Program Material For Four-Channel Playback.* Jour. Audio Engr. Soc., Vol. 19, No. 4, April 1971 (pp 262–266).

7. Eargle, John. *Stereo/Mono Disc Compatibility: A Survey of the Problems.* Jour. Audio Engr. Soc. Vol. 17, No. 3, June 1969 (pp 267–281).

8. Cunningham, James and Bruce Swedien. *Thoughts on Quad Mixing.* Recording Engr./Prod., Vol. 5, No. 5, October 1974 (pp 47–53).

9. Revill, Adrian. *The Tempest.* Studio Sound, Vol. 16, No. 8, August 1974 (pp 32–34).

10. Walker, Saul and Paul Buff. *A Practical Approach To Recording Studio Automation.* presented at the 43rd convention of the Audio Engr. Soc., September 1972 (Preprint 901E-7).

11. Jones, Wayne. *Automation as Applied to the Mixdown Process.* presented at the 41st convention of the Audio Engr. Soc., October 1971 (Preprint 817L-3).

12. Marrone, Joseph. *How About Some Automation?* **db.** The Sound Engineering Magazine, Vol. 6, No. 8, August/September 1972 (pp 38–41).

13. Jung, Walter. *Automating the Audio Control Function.* **db.** The Sound Engineering Magazine, Vol. 6, No. 4, April 1972 (pp 24, 26–28, 30, 32); Vol. 6, No. 5, (pp 26, 28, 30–33); Vol. 6, No. 8 (pp 46–48, 50–51); Vol. 6, No. 10 (pp 30–35).

Multitrack in Action

The elaborate equipment we have considered obviously has great potential and the multitrack techniques are interesting, but just how is everything put together to do a particular recording job? In this chapter we are going to follow an actual recording and mixdown session in detail. Concentrating more on edification than amazement, we have selected a small unknown musical group, typical of untold thousands around the world, struggling for recognition, and we shall use minimum equipment: a small console (Tascam Model 10), an 8-track recorder (Tascam Series 70), and a 2-track master recorder (Teac Model 3300S-2T) using ¼ in. tape. We are deeply indebted to officials of Teac Corporation of America and to Tascam for their cooperation in describing this recording session. The following is adapted from descriptions of several actual recording and mixdown sessions of this musical group at the Tascam Studio, directed by company personnel.

The musical group was composed of five instrumentalists and a female vocalist. The leader of the group played three keyboards, a Hohner electric clavinet, a Fender/Rhodes electric piano, and an ARP string synthesizer. The leader also sang lead and harmony vocals. In addition, there was a drummer, an electric bass player, a conga drums player, and an electric guitarist.

The recording took place in a 20 by 30 ft studio with the musicians arranged as shown in Fig. 9-1. The drummer with his kit of drums was positioned in the far corner of the room, surrounded by heavy 4 by 8 ft baffles made of 1½ in. solid wood cores that were covered on both sides with fiberglass and fabric. An opening allowed the drummer to have visual contact with the group leader on keyboards in the diagonal corner of

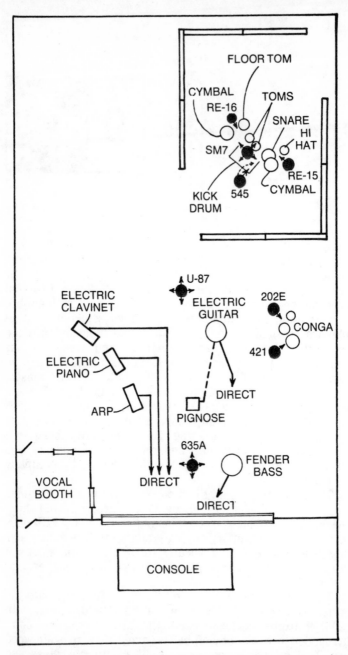

Fig. 9-1. Arrangement of musicians and microphone placement for the recording session described in detail in the text.

the room. The drummer had a full kit: kick drum, hi-hat, two cymbals, snare drum, and three toms (one a floor type). The conga drums, electric guitar, and Fender bass were placed between the drums and the control room window as shown. As the conga drums are not as loud as the other drums, they were placed carefully to avoid leakage into their mikes from the drum set.

The vocal booth is telephone booth size with two windows giving visual contact with the leader and the control room. The vocal booth was not used for the recording of the vocal tracks. They were all overdubbed because the sound from the booth was too dead. The booth *was* used, however, during the ballad. The group was experiencing difficulty getting good basic tracks because they were not hearing the vocals the way they did in live performances. The female lead vocalist was placed in the booth where she sang live for the benefit of the performers over the cue bus. Her voice was not recorded while in the booth, but overdubbed later from the center of the studio. The Neumann U-87 microphone was adjusted to give a cardioid pattern and placed in the center of the room. Vocalists were positioned about 12 in. from this mike. A slight amount of room reverberation helped the effect.

MICROPHONE PLACEMENT

For voice recording, four basic tracks would handle the first pass. The only instruments picked up acoustically were the drums and congas. All the electrical instruments were picked up at their preamp outputs and fed directly into the Tascam board. Of course, these instrumentalists had to have headphone feeds to hear themselves. Picking up the electrical instruments with microphones in front of their loudspeakers often runs into noise problems, and, of course, leakage to other mikes can also occur. In spite of this, a mike positioned before an instrument loudspeaker often gives the sound desired.

Four microphones were used on drums. The front head of the bass drum was removed and a horse blanket was packed inside to absorb some of the fundamental. A Shure 545 microphone placed inside the drum provided the necessary bass drum signal for the track. The stereo pickup of the drum set was accomplished with Electro-Voice RE-16 and RE-15 cardioids. They were mounted close and pointed downward, one over the toms and the other over the hi-hat. In addition, a Shure SM-7 cardioid was positioned directly in front of the drummer himself.

Two microphones set for a cardioid pattern (about 5 ft above the floor and in a stereo arrangement) picked up the congas. One mike was an AKG 202E and the other was a Sennheiser 421. Aimed downward, they caught the direct sound from the skins as well as the bounce from the floor. (This arrangement, incidentally, was adopted only after extensive experimentation with mikes closer to the floor.)

These were the only microphones used—except for the electric guitar, which was overdubbed. On one of the songs, the guitar was fed into a *pignose bass* amplifier, which has a very small loudspeaker and considerable distortion—but the sound is distinctive and generally sought after. A direct feed from the pignose preamp was augmented by a microphone pickup on the pignose loudspeaker. The microphone used for this was an omnidirectional Electro-Voice 635A mounted about 10 ft away from the pignose.

CHANNEL ASSIGNMENT

Now let's go into the control room and see how the instruments were assigned to the various channels of the console. The standard Model 10 Tascam board has 8 input channels, although there is room for 12 on special order, and 12 more can be added in an expander desk. This one had 12 channels. Table 9-1 lists the channel assignment used. There was really no powerful rationale behind these decisions. It was more a matter of habit the mixer had developed over the years. For instance, he was accustomed to having the drums on the left of the console and other instruments on the right. The kick drum was fed to channel 1 and the overhead drum microphone to channel 2. Channels 3 and 4 were fed by the stereo drum microphones, the audience left mike over the

Table 9-1. Console Channel Assignment

Channel	Instruments
1	Drum—kick
2	Drums—overhead
3	Drums—left stereo (toms)
4	Drums—right stereo (snare)
5	Conga drums—left stereo
6	Conga drums—right stereo
7	Electric piano
8	Clavinet
9	Bass

Table 9-2. Magnetic Recorder Track Assignment

Track	Assigned to
1	Keyboards, electric piano, clavinet (mono)
2	Drums & conga (stereo left)
3	Bass (mono)
4	Drums & conga (stereo right)
5	Vocals (double tracked)
6	Keyboard (melodic lead)
7	Vocals (double tracked)
8	Guitar

toms to channel 3, the audience right mike over the snare to channel 4. Channels 5 and 6 handled the stereo pickup of the conga drums, audience left on channel 5 and right on channel 6. The first six channels were given over to getting a good drum sound, a very important part of modern rock music. The electric piano, clavinet, and bass fed individually into console channels 7, 8, and 9. The vocals and guitar were overdubbed, so they came a bit later.

There are some good reasons behind the recorded track assignments of Table 9-2, and one of them is tape stretch resulting from repeated playback. It is not unusual for a master recording to be played several hundred times before work is completed, resulting in tape deformation which chiefly affects the outside tracks (tracks 1 and 8 or 16). *The edges are not used for solo or other vocals and, if possible, not for instruments having appreciable energy at the higher frequencies.*

As the magnetic recorder was an 8-track unit, a bit of premixing was necessary. The drum channels were premixed in stereo to tracks 2 and 4 as shown in Table 9-2. The electric piano and clavinet were mixed onto track 1, and the bass put on track 3. The high-priority drum tracks were separated to reduce crosstalk between them which would tend to defeat the stereo effect and were kept off tracks 1 and 8. Because there is less high-frequency information on the electric piano and clavinet, they were assigned to track 1. The same is true of the guitar which was placed on track 8 in overdub. The keyboard melodic lead and vocals filled tracks 5, 6, and 7 in overdub

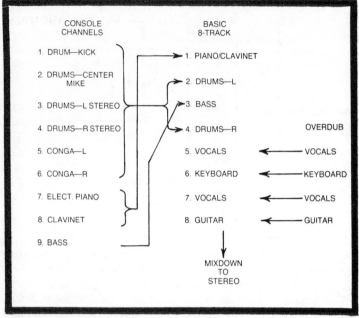

Fig. 9-2. Overall plan for the recording of the basic 8-track tape which was later mixed down to stereo.

sessions. Figure 9-2 shows clearly the premixing necessary to work within the limitations of the available equipment. If a 16-track recorder and a board with enough channels had been available, all the drum mixing would have been deferred to the mixdown session.

The overdub of the guitar was made with only the guitarist in the studio—the rest of the band was in the control room kibitzing. Of course, this gave perfect separation of the guitar track, which allowed complete freedom in manipulating it later. The keyboard melodic lead was overdubbed next, followed by the vocals.

The lead and harmony vocals were premixed and double-tracked in the overdub. This means that after the first vocal track was recorded on track 7 the tape was rewound and the vocalist recorded the same thing on track 5. The vocalist tried hard to make the second pass just like the first one, but of course there were slight differences which, when the tracks were mixed together, gave the resulting vocal track a desirable quality of fullness which many people enjoy. This is a different and valid artistic use of multitrack recording but is less popular now than a few years ago.

We must go back and pick up a bit of detail on what was done in the line of limiting and equalizing as the basic tracks were recorded. Actually, it is best to use the absolute minimum of limiting and equalization in the original recording because this retains the greatest freedom for mixdown manipulations to get the desired effect. Also, if the original tracks are "pure," one can do another and different type of mixdown later if desired. In the present case, very little limiting of the bass track was needed, about 4 to 6 dB only, because the bass player was a very consistent artist. The recording engineer tried limiting on drums but finally decided to ride gain instead, especially during high-level passages. A 12 dB per octave rolloff at 10 kHz was applied to the kick drum; the signal was given a 3 to 5 dB boost at 3 kHz to enhance harmonics.

There was a particular notch-loss problem with that drum, so the output was given an 8 to 10 dB boost at 90 Hz. It would have been preferable to do this at 70 or 80 Hz, but 90 Hz is as low as the equalizer went. Peaking at 90 Hz, however, gave a good boost at 70 Hz also.

For the hi-hat and the snare drum a 3 dB boost was added at 10 kHz and another 3 to 5 dB at 5 kHz for more presence; everything below 200 Hz was rolled off at 12 dB per octave. This helped the noise a bit and kept the toms from coming through that mike. More or less the opposite was done on the toms by pulling them down about 5 dB at 10 kHz.

The overhead drum mike was left flat except for a minor cut of the lows. There was a little problem on the congas getting a satisfactory location for the two mikes; but once solved, the highs were cut a bit on the mike next to the large conga and the highs were boosted on the small conga. The drummer played on the rim of the smaller one; and to enhance this sound a 3 to 5 dB boost was applied at 10 kHz. The midrange was kept flat on both.

On the clavinet and electric guitar a hiss problem originated in the vacuum-tube preamplifiers. This was greatly helped by applying a 12 dB per octave rolloff at 10 kHz. As a compromise, a dip was applied at 5 kHz. This is the sort of price one pays using instrument preamplifiers.

CUE MIXES

Musicians on drum, conga, keyboard, and bass required foldback, which proved troublesome because each wanted a different mix in his headphones. As there were only two cue mixes, it took a bit of fast talking. The drummer and the conga players were finally convinced that they could use the same

Table 9-3. Cue Mix Content

Foldback	Cue Mix 1 (low drums & conga, full keyboard & bass)	Cue Mix 2 (Full drums, conga & bass, low keyboard)
Drums	x	
Conga	x	
Keyboard		x
Bass		x

mix with minimum level drums because they could hear themselves without headphones. The keyboard and bass players were also convinced that they could function on the same mix (Table 9-3), with full drums and bass and lower levels on their own instruments. In this particular group, everyone wanted to hear the bass.

The Model 10 Tascam board can provide a cue mix only by using the echo send (later models correct this). For this reason two Tascam boards were used, one for recording and the other for providing the two required cue mixes. The echo bus of the second board provided one cue mix, and the program bus provided the other.

MIXDOWN

The basic 8-track recording was now in hand and the next job was to play the 8 tracks into the board for the stereo mixdown. For this the tracks were assigned to correspondingly numbered channels for convenience. The mixdown fader and panpot settings required much experimentation that is too complicated to explain, other than that it was a search for the best way to achieve the mental image of the finished product. The outcome of this is illustrated in Fig. 9-3 as far as the panpots and faders are concerned.

First, the bass panpot was put on dead center for channel 3. The double-tracked vocals were split, channel 7 to audience left, channel 5 to the right. This may seem strange, spreading vocals across the stage, but the effect gives greater fullness. Actually, if the two vocal tracks were exactly synchronized, they would apppear to come from the center with the panpot and fader settings shown. The differences would tend to come from left or right, creating a desirable effect. The keyboard

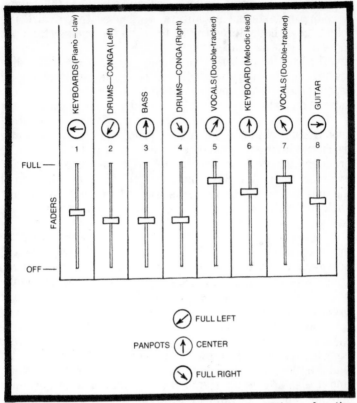

Fig. 9-3. Approximate fader and panpot settings for the mixdown.

and the melodic lead were placed in dead center with the bass. The guitar was moved over to about 3 o'clock right and the other keyboard to 9 o'clock left. The left drums and conga track would, of course, be full left, and the right drums and conga full right. There were changes from song to song, but this is basically the arrangement with this group.

The fader positions of Fig. 9-3 are approximately those which gave the best balance. Of course, all this was determined by ear—not by VU meter readings or fader settings. It was a vocal selection, so the vocals were dominant, followed by the melodic lead keyboard. The guitar and other keyboards were at about the same setting. Drums and bass had the lowest fader settings, but of course they came through well because of their percussive character.

Some reverberation was applied to the vocals, the amount depending upon the song. A bit of reverberation was also applied to the lead melody keyboard, and a touch to the drums. A couple of external equalizers were used to equalize the midrange of both stereo tracks to make the vocals stand out more from the accompaniment.

The stereo master is now ready to be sent to the disc mastering laboratory where it can be cut in quadraphony by sending the stereo through a matrix encoder. To help with the noise, it could have been Dolby-encoded in the mixdown.

Overdubbing may have to be done over and over before a good track is obtained. There is no problem in erasing a faulty track and laying down another in its place. In fact, it is quite practical to punch in and out to redo just a section. For example, the guitar player hit a bad chord at the very end of the song, yet everyone was pleased up to that point. By running back a few bars to a convenient place, he rerecorded the ending.

On one of the songs the piano player came in intermittently with little riffs. A Teac AN180 Dolby-B processor was used to achieve a noise-gate effect so that the track was essentially shut off when there was no signal, achieving a 10 dB reduction of tape hiss on that track in the process.

One of the pitfalls for the novice results from having monitor levels set high and program faders set low. It's very easy to do that. This creates an imbalance between the meter and the monitor, it affects the frequency response of the ear, and it also affects the signal-to-noise ratio of what goes onto the tape, and the inevitable result is a hissy tape.

RECORDING OUTSIDE THE STUDIO

Sooner or later—usually sooner—a recording organization is requested to record outside the confines of the studio. Refusing such requests will be a step in the direction of building a reputation for specialization the studio may ill afford. This matter of defining the limits of the services offered requires serious consideration of just what type of jobs can be done best with equipment and talent available. Once decided, it is best to keep within the limits set and build a solid reputation in that arena. It is also possible to build another kind of reputation the studio can ill afford—one of not doing good work—which might come as a result of doing a remote job with makeshift equipment.

The Transportable Approach

Recording jobs outside the studio can range from recording conference speakers requiring nothing more than a consumer-type tape recorder and a microphone or two to recording in multitrack a full symphony orchestra in a big music hall or an open-air rock concert. The smaller jobs may easily be handled on a quality basis with a good portable recorder, a few microphones, and a portable mixer. The big jobs may demand a full-sized console and multitrack recorders and all the multitude of microphones, stands, and other ancillary equipment. Are you willing to rent a truck and move your precious console and other equipment to the recording site? Modern consoles are rather bulky and difficult to move in the physical sense; but fortunately they are beautifully self-contained. And if you have the hundreds and hundreds of feet of microphone cable, it is simple to plug in the mikes, plug in the power cord, and put it into basic operation in minimum time. Some of the smaller desks are more adaptable to this service than the larger ones, but moving the large one is a possibility, though a rather overwhelming one. Whether such a move is practical may depend upon the length of the recording engagement, the size of the fee, and impact on the studio schedule.

The Mobile Approach

The studio organization may decide that there is enough recording work available outside the studio to justify investing in a mobile recording facility. This is a fully equipped control room which can be driven to the location and placed in operation immediately. Further, when the doors are locked, the equipment is easily secured against vandalism, theft, or unauthorized tampering. Such mobile equipment has taken many different forms and is quite widely used in Europe and America.

Trailers may be used as mobile recording control rooms. These may range all the way from the home-type recreation vehicle to giant trucking type trailers or semitrailers hauled to a location by a large automotive tractor which may be uncoupled and driven away. Such trailers may be as much as 30 ft in length and 8 ft wide.

Trucks can provide space for a control room and offer the advantage of having their own motive power available on a moment's notice. Again, these can range from a VW van to the huge monsters, depending on the space required and the size of the budget.

Containers of the type used on shipboard and transported on land by flatbed truck or by rail have even been used for housing a transportable multichannel console and multitrack recording equipment. A company in England doing many field recording jobs on the continent found the container to be an excellent solution because of the way it simplifies customs inspections. Once inspected by officials in England, the container is sealed, shipped to the continent where it is inspected once more, and that's it.

Power Supply

Electric power generally can be obtained by tapping into local service power points, but in recording an outdoor event there may be no power available. Gasoline engine driven generators are the answer for such situations. If mounted on a separate small trailer, the generator can be located a distance from the recording van to reduce the noise. If mounted close—for instance, behind the driver's cab and ahead of the main van—great precautions must be taken to control the noise level. This problem should be approached through (1) proper resilient mountings of the generator to the frame, (2) enclosing the generator and muffling the engine properly, and (3) giving extra attention to the sound transmission loss of the adjacent van wall. Of course, such noise would be of primary concern in monitoring, as the microphones would normally be at some distance from the recording van.

Visual and Audio Communication

There is no convenient observation window in remote recording allowing the recording engineer to see what is happening around the mikes. Closed-circuit television is the almost universal solution to this problem because visual cues are often very important to a smoothly run session. Often several TV cameras and monitors are required. An elementary intercom facility may be needed between van and stage, and a radio link would allow the technician setting up the mikes great freedom in moving about without a tangle of trailing wires.

Acoustic Treatment of Mobile Vans

Acoustic compromises are the name of the game in location recording. There are compromises on the microphone end trying to get adequate track separation under less than adequate circumstances. The necessity of operating a sound reinforcement system for the benefit of the audience during the recording is a very common headache, as this type of recording is often associated with big name performers, large halls, and big audiences. There is a corresponding compromise in the van control room. First, it's too small (or at least too narrow), and this moves the normal-mode problem to ever higher frequencies. Second, the dimensions are probably controlled by the highway authorities, by the equipment available, or by the budget, rather than by acoustical factors. Under such circumstances it is understandable if the design engineer throws up his hands and specifies overly absorbent surfaces. The effect of poor listening acoustics would appear primarily in the reduced sharpness of the operator in catching the more obscure problems.

Microphone Cable Runs

For full-scale recording jobs, cable runs of at least 300 ft would undoubtedly be required—and twice this at times. Long runs mean great exposure to hum and noise pickup, and care must be exercised to make sure that all circuits are properly balanced and properly grounded and, of course, of suitable low impedance.

Recording Van as Studio Expansion

Some recording organizations have helped to justify the great cost of a big mobile recording unit by using it as an extra mixdown room as it is parked on the back lot. If the equipment is adequate, and it generally would be, about the only objection would be based on acoustic compromise. It is admitted that all too often battles between dollars and acoustics result in acoustics losing out. Once the customer begins to notice, however, this might prove to be poor economy. With the limitations of a mobile studio clearly in mind, its use as a mixdown room is a possibility. Surely, the acoustic faults for mixdown are the same as those in recording; and if the mixdown is of the material recorded in the mobile unit, the acoustic problems are compounded.

10 Locating and Laying-out the Studio

It is entirely possible that selecting the right location for a multitrack recording studio might be more important to the financial success of the venture than all other factors combined. In deciding on a site, the potential studio operator comes face to face with the problem of catching the eye of future customers. Should he go where the potential customers are? Or do they expect him to be somewhere else? Such intangibles as image and customer expectations and solid tangibles such as environmental noise, zoning restrictions, and cost should be carefully weighed in the vital decision of locating a studio.

LOCATION

A very fundamental factor in locating any business is to make it convenient and attractive to the potential customers. In the case of the recording studio, the potential customers are vocal and instrumental artists and musical groups of various kinds. How will the projected new studio offer new and different enticements to them? Location of the studio in a specific part of town may be important. Distressing as this may seem to the technically oriented, the "right" neighborhood may be just as significant to an artist as a blossoming array of knobs on a console! A manifestation of the generation gap may creep in at this point. A 50-year-old establishmentarianist businessman contemplating investing in a recording studio may have difficulty imagining what appeals to modern 20-year-old musicians. He undoubtedly needs the help of one who can speak the language and move freely among potential customers—someone who can serve as a sort of cross cultural consultant. It is significant that so many of the successful recording studios have been established and are being run by young pop musicians.

BUILD, BUY, OR LEASE?

There is something intrinsically satisfying about the idea of constructing a studio building that exactly meets specific needs. But it is an expensive, drawn out, messy process. Someone intimately involved in the studio operation and one thoroughly informed in the technical and operational aspects of recording must devote endless hours in layout and planning before an architect can take over. There is no doubt, however, that having a studio complex built to one's specifications yields the best fit to individual requirements.

Buying an existing building and adapting it to serve as a studio is another obvious way to go. Here, the crux of the matter is how well the various available structures fit the particular requirements of a recording studio. For example, a building built to serve for manufacturing may have thin walls and roof which offer little isolation from outside noises or protection for neighbors from sounds generated within. The cost of patching up such deficiencies *can be prohibitive*. If the location, integrity of structure, and layout are reasonably satisfactory, buying an existing building and adapting it to studio use may prove to be economically and functionally advantageous.

Short-term renting is normally ruled out because of the cost of renovation and adaptation. As these costs must be included in the operating expenses and spread over a number of years, a lease agreement is required to guarantee occupancy. If leasing a separate building is under consideration, then all the *buy* factors apply equally (adaptability to studio use, cost of alterations, etc.). Often, however, the leasing arrangement applies to space in a building occupied by others—and this opens up a whole new set of potential problems.

Of course, the choice between building, buying, or leasing will ultimately be decided on the dual basis of suitability and cost.

If you decide to build your own structure, retaining a good architect to draw up the plans and specifications and to supervise construction usually proves to be the cheapest and best in the long run. Often there is the temptation to engage only a building contractor for the entire job. While the abilities of these gentlemen are vital to a construction or remodeling project, they usually fall short in the esthetic and creative aspects of planning and design. And remember, appearance and impressions are important to the artists who will be the customers. Also, there will be no one but the local building inspector to check on the contractor's work. Bear in mind that he checks only for conformity to local codes and not to contract specifications.

Assuming comparable suitability, cost becomes everything. This must be approached on the basis of cost per year of service. The high first cost of building a new structure or buying an existing one must be reduced by the estimated resale value at the end of a given number of years and the cost per year determined from the difference. Alteration and adaptation costs are different in that the probability of all those special studio walls being of value to the next tenant is extremely remote. They will probably be considered only as an expense for removal. Thus, the prorated annual cost in building or buying must be compared with competing lease costs to decide between them. To this annual cost must then be added the prorated annual cost of making the space into a studio complex to find the total cost of housing the projected recording activity.

TAXES AND ZONING

One never escapes property taxes. They will be paid directly in the buy or build cases, and indirectly in the lease case. Sensitivity to high or low tax areas should be a factor in site selection.

Local government enters the entrepreneur's picture in many ways. Property taxes go along with ownership of site and studio. A local business license is required and registration with the local sales tax authority is often a separate transaction requiring periodic reporting and remitting.

The establishment of new quarters for your recording activity will involve local government in many other ways. Before taking definite steps toward acquiring property, you should be certain that the area is zoned for your type of activity and that area development plans for the future will not change this. It is not enough to know in your heart that you will not be creating a nuisance which might give rise to complaints later; the important thing is, are you legally entitled to be there? These things can seem very arbitrary and unfair but must be followed unless you are prepared to risk your investment in a fight with local officials from city hall.

TECHNICAL SERVICES

The dramatic increase in the number of recording studios in recent years has spawned a number of studio consulting organizations. Any one of these would be very glad to take over all your headaches in getting established—for a price. They offer the advantage of complete familiarity with the specialized field of recording studios, several of them having guided the establishment of dozens of studios (called *turnkey* installations) from the ground up. They will undertake studies of feasibility, site location, site survey, will supervise alteration, construction, acoustic treatment, and will be very glad to equip the studio completely for you. If you are inexperienced in the field, this could be a satisfactory solution to your problem.

Precautions should be taken before entering into an agreement with such a studio consulting firm. Because they usually represent certain lines of consoles, recorders, and other major equipment, one can expect an extolling of the virtues of this particular equipment, the specifications of which should be independently checked out. It would be advisable to consult personally with former customers of the firm to ascertain satisfaction and to evaluate the firm's ability to deliver. Fortunate is the one considering a sizable investment in a new studio who already has engaged a competent audio engineer to help in such investigations.

Even after the agonies of installation are crowned with the laurels of initial success, there is the little matter of keeping the production wheels turning. This can be a bit tricky with the sophisticated equipment that is the heart of a modern multitrack recording studio. As competent as your technical staff may be, outside consultant help and access to manufacturers' representatives is sometimes needed and time is the important element. At such times the local availability of technical services may be of great value. Further, a ready source of equipment and supplies such as tape, normally found in the larger cities, may be an economic advantage.

ACOUSTIC FACTORS

Outside noise such as that produced by aircraft may affect the recording being made in the studio. This is the time-honored way of considering noise and its effect on one particular kind of sound recording. If you are considering the recording of rock music, things may be rather reversed. Not that an earthbound band inside a studio could disturb passengers in a high-flying aircraft, but the band might be considered a class A nuisance by a next-door psychiatrist with a patient on the couch! In a later chapter we consider the kind of studio walls that will attenuate sound going in either direction; here, we are simply calling attention to both types of potential noise problems: (a) the outside noise which might intrude upon soft passages being recorded in the studio, and (b) high-level sounds from within which might escape sufficiently to bother neighbors. Both these factors should be considered in locating a recording studio, because the success of a recording studio may well depend upon taking a wide variety of recording jobs.

We are becoming more aware of noise pollution of our environment from the health standpoint; it is a fact that noise levels vary widely from point to point. Anyone would hesitate to establish a recording studio next to the tumblemill of an iron foundry or under an aircraft takeoff pattern of a busy airport. But how about traffic noises due to automobiles, large trucks, and screaming light motorcycles? It is just possible that the air-conditioning equipment or the offset press in the office next door may be more troublesome than any outside sound.

OTHER CONSIDERATIONS

The client is a bit disheveled and flushed as he arrives to arrange for a recording session. He has just spent the last 20 minutes looking for a parking space. The one he finally found is six blocks away. And if he is a bit on edge this time as he arrives, how is he going to react to a citation for incorrect or overtime parking?

Your prospective clients expect comfort, convenience, and pleasant surroundings. Even if they are unreasonable in such expectations, is it the place of the recording studio operator to try to reeducate his clients? Where big decisions can hinge on small and intangible things, the answer must be a resounding *no*! The wise studio operator makes every effort to put the client at ease and in a good frame of mind. If this requires validation of a parking ticket, so be it. This is a legitimate expense of operation and one which might yield big returns.

Convenient parking facilities are no less important than an attractive entranceway, convenient and clean lavatory facilities, and a never-empty coffeepot. In fact, the coffeepot is important for more than initial impressions—all-night sessions pivot on black coffee. The point to all this is that it is not enough to be just an all-business organization housed in any old structure. The message of your ability to do a good recording job reaches the client through such humble means as the parking lot, the clean lavatory, and the coffeepot. The communication specialists have a term for it—*nonverbal communication*—and they claim that an amazing 65% of the face-to-face messages we receive daily come through other than verbal channels. Realizing this, the carpet on the floor and the neat, friendly receptionist who is reasonably civilized in the way she chews her gum can all take on higher meaning and value.

A word of warning regarding the projection of too rich an image: What this recording job is going to cost is very probably uppermost in the mind of the prospective client as he first visits your studio. If he sinks halfway to his knees in carpet and sees that the studio manager's desk is solid Thai rosewood and as big as a tennis table, he may well wonder, "Wow! Am I going to have to pay for that, too?" The line between the image of competence and the image of opulence isn't all that fine and difficult to follow, however. Being aware of the distinction is really a sufficient guide.

All the foregoing emphasizes the importance of a very careful survey of the potential studio site, whether the buy, build, or lease path is followed. Here is a checklist to guide such a survey.

SITE SURVEY CHECKLIST

1. *Business factors*
   - Cost comparison of building. buying. leasing
     - —capital investment
     - —operating expense
   - Local government requirements
   - Proximity to talent pool
   - Availability of technical services and supplies
2. *Emotional factors*
   - The neighborhood
   - Studio facilities
3. *Acoustic factors*
   - Outside noise which might interfere with recording
     - —aircraft
     - —traffic
     - —industrial
   - Potential noise problems within the building
     - —air conditioners
     - —elevators
     - —office machines
   - Neighbors sensitive to noise
     - —schools
     - —hospitals
     - —professional offices
     - —homes

Laying Out The Studio Complex

Studio layout is a jigsaw puzzle. Each specific function of the studio complex can be represented by a scaled-down paper cutout of certain shape and size; the cutouts can be shuffled around to provide the desired interrelationship of functions to give at least a rough starting point to overcome that awful *what-do-we-do-first* feeling. Once this first approximation is down on paper, refinements will naturally come as a result of further thinking about details of operation. The greater the experience of the persons doing this, the more functional will be the resulting layout. The information offered here is intended to serve as a guide in this intricate process and a reminder of the thousand-and-one things going into the makeup of a recording studio.

How Big Is A Studio?

The greatest uncertainty surrounds the number and size of the studios to be incorporated in the plan. A very high percentage of the floor area, and hence the cost, will be in studios, and yet right here specific figures are hard to come by. One approach is to go to the reference books. For example, in discussing recording studios and motion picture scoring stages, Rettinger says[1] that the number of musical instruments N is room volume V in cu ft, by the expression

$$N = 0.0021V^{0.855}$$

Expressed in another form, the volume V is approximately dependent on the number of instruments N to be accommodated according to :

$$V = 1350N^{1.17}$$

We recognize immediately that the studio volume required depends upon the type of recording to be done. The volume indicated by the equation might be too small because of the need for acoustic isolation between performers or groups of performers.

Low-frequency resonance conditions in small studios have been mentioned and are once more relevant. Although a speech studio need have only a very small volume to accommodate a single narrator physically, very small studios are generally difficult to handle acoustically. Gilford, with his wealth of experience as head of the acoustic section of the British Broadcasting Corporation for over two decades, says of *talk* studios, of which BBC has several hundred:

> Volumes from 1500 to 4000 cu ft are generally satisfactory, bad colorations being difficult to avoid in studios below this range, and larger studios giving insufficient advantage to justify the increased expense of construction and treatment.[2]

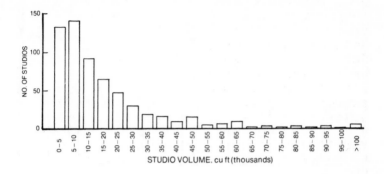

Fig. 10-1. Studio size distribution of the 602 recording companies in the U.S. listed in the 1973 Billboard International Directory of Recording Studios.

Another approach to the problem of studio size is to find out the size of recording studios already built and in service. The Billboard International Directory of Recording Studios[3] gives some information on this subject. A study based on studio listings in this directory yields some very interesting information. In Fig. 10-1 the distribution of sizes of some 602 U.S. recording studios is shown in bar-chart form. If a recording company has more than one studio, only the largest is included. Figure 10-1 reinforces our general conviction that there are more small studios than large ones, but adds the definite information that 60% of the recording organizations in the U.S. have studios (their largest if they have more than one) 15,000 cu ft or less in volume and only 6.8% have studios 50,000 cu ft or larger. Figure 10-2 expands the data for the first four bars of Fig. 10-1 to show how the smaller studios are distributed in size. These small studios are important to us because, for better or worse, such a great proportion of the recording business going on is done in them.

In Fig. 10-3 each spot represents one of the 535 U.S. recording companies listing the required data in the Billboard directory. If a company has more than one studio (and about a third of them do), again only the largest studio is represented in this plot. For example, the listing of the first studio in the directory says: *Size 40 × 30, height 14 accommodates 25.* The volume ($40 \times 30 \times 14 = 16,800$ cu ft) is then plotted against the 25 people or instruments accommodated, resulting in one point on the graph of Fig. 10-3.

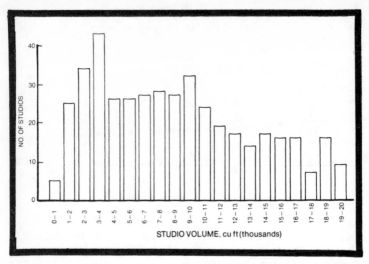

Fig. 10-2. Expansion of the data of Fig. 10-1 for studios 20,000 cu ft and smaller.

There is tremendous variability in estimates from different recording companies as to just how many people a studio of a given size will accommodate. Part of this variability is undoubtedly a reflection of the different types of recording jobs done. When one operator says 100 people can be accommodated in a 50 × 30 × 12 ft (21,000 cu ft) studio, this doesn't sound too unreasonable; but how about the one who says his 21,000 cu ft studio accommodates only 25? Dozens of operators having studios ranging in size from 15 × 10 × 9 ft (1350 cu ft) to 70 × 40 × 20 ft (56,000 cu ft) say their studios accommodate 10 persons.

In Fig. 10-3 the smooth curve labeled N is plotted from the equation. Practically all the points from the industry lie above this curve. Another curve, labeled $3N$, representing three times the values obtained from the equation, is also drawn in. This comes much closer to running down through the center of the industry spread. This is not to say that $3N$ curve is right, but only that it conforms somewhat better to the claims of the recording industry. If all we want to do is to build a studio like the others, following the $3N$ curve would be a rough approximation of the average of what the other recording companies are doing in regard to studio volume required to accommodate a given number of performers.

In all fairness, we should be aware that a listing in Billboard's directory is part of the recording company's sales promotion. The manager may claim that 20 performers can be accommodated in a studio 20 × 18 × 8 ft (2880 cu ft) or 18 sq ft per person, but it may be a slight exaggeration, hopefully in the direction of getting more business. This is the size of a not-so-large living room and the 8 ft ceiling would be a real problem. It is also possible that the point on the other end of the 20-person spread of Fig. 10-3 is too conservative: 70 × 48 × 20 ft (67,200 cu ft) or 3360 cu ft per person.

Different groups being recorded and different recording techniques used may make both of these extremes seem reasonable. If a 20-voice choir were recorded with one mike, maybe the 20 by 18 ft studio could serve in a pinch. On the other hand, a 19-piece feature band plus a vocal soloist recorded on 24 channels could put a strain on the 70 by 48 ft studio to achieve the necessary separation. This illustrates the necessity of determining studio size by a careful consideration of the types of jobs to be done and the recording techniques to be employed as well as the size of the groups to be accommodated.

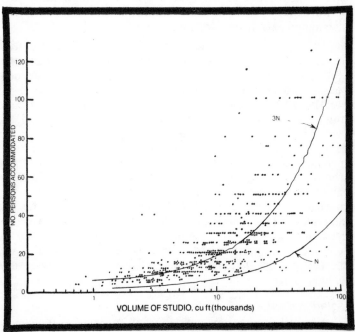

Fig. 10-3. Number of artists accommodated related to studio size. Each spot represents the largest studio of the 535 recording studios listing studio capacity. (Data from listings in the 1973 Billboard International Directory of Recording Studios.)

Table 10-1. Studio Data[3], 673 Listed Companies

Category	No. Firms	% of total
One studio	425	63
Two studios	151	23
Three studios	48	7
Four or more studios	13	2
No studios (those firms specializing in remotes, mastering, editing, or mixing or no data given)	36	5
Total company listings	673	100

How Many Studios?

The Billboard directory also gives us information on American practice in regard to the number of studios needed or affordable. There are 673 listings in the 1973 directory, distributed as indicated in Table 10-1.

Often the second studio is a small voice studio or isolation booth, but if designated for operational tasks such as mixdown, editing, dubbing, or mastering, it is included not as a studio but as a production workroom.

The ultimate decision on the number of studios to build must grow out of a projection of the type of work expected, the recording techniques to be used, the size of the largest groups to be recorded, as well as economic factors such as estimated income, operating costs, etc. It is strongly urged that those who must start small because of modest capital should avoid the tiny cubicle studios which leave their acoustical coloration mark on every recording. The too small rooms and the too low ceilings will be a continual source of problems with no simple economic solution. It is suggested that the BBC lead be followed and studios of less than 1500 cu ft be avoided. Even these, and those up to 3000 or 4000 cu ft will need very careful acoustical treatment.

Isolation Booths

Vocalists may not have enough acoustical power to compete with the instruments or it may be considered desirable for the vocal track to be essentially free of background accompaniment for maximum flexibility. An isolation booth may be the answer. Such a booth is simply a small studio with all the acoustical demands of any studio, preferably located adjacent to the larger studio in which the accompanying musicians are working and provided with windows for visual contact. Such rooms should have a volume of at least 1500 cu ft to avoid excessive problems with room resonances.

A *drumbox* is sometimes used to give adequate separation between the percussion section and the other instruments in the main studio. Possibly drums require less attention to acoustic performance of the room than voice, but good acoustics will not hurt the drum track and the flexibility thus achieved may fully justify the treatment. Visual contact is not always necessary.

Control Rooms

How many rooms are required? How large must they be? Each studio should have its own control room. The idea of sharing a control room with two studios will work only if the two studios are used one at a time. If the mixing staff is one person, or business is so poor that the above condition is guaranteed, should a shared control room be considered at all? Who knows? Maybe business will improve and you will need another operator and it will be necessary to use both studios simultaneously. If you don't expect much business, forget about the small studio. A postrecorded narration or commentary can be recorded in the larger room quite satisfactorily (unless it is really cavernous) and, if business thrives, not having a small studio will help you resist the temptation of putting too many performers in it. There may be an exception to this if the small studio is used as an isolation booth along with the larger studio.

Even with duplicated consoles in the shared control room, the problem is one of listening on loudspeakers to what is being recorded. This can be done on only one program source at a time in a given control room, obviously.

Although sharing a control room with two recording studios has its great disadvantages, there is a type of sharing that makes good sense. This is the case of the smaller recording company which can afford only one console to be used for both recording and mixdown. If the control room listening environment is good enough for recording, it will be quite satisfactory for mixdown and final checking of mixed tracks. The scheduling problem remains, but as long as there are 24 hours in the day and the coffee holds out, the job can be completed.

Size of the control room is important from the standpoint of equipment and personnel as well. Today's consoles tend to

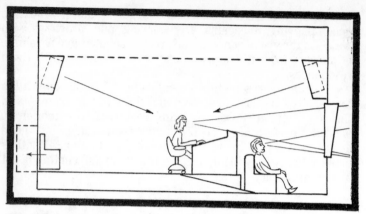

Fig. 10-4. An efficient control room arrangement which is widely used.

be large. Many consoles require access to the back face, which means they should not be jammed up against the double glass window wall. Placing a console some distance from this wall allows better loudspeaker—mixer orientation, placing them in front of the mixer rather than overhead or to the side. Rear loudspeakers also take up space if required for monitoring a 4-channel mixdown. Multitrack recorders take their share of floor space as do racks for peripheral equipment.

The geometric shape of the control room should be adjusted to distribute normal-mode resonances in an optimum manner, as mentioned in Chapter 6. The physical relationship of the control room to the studio it serves is important. A very effective control room arrangement is pictured in Fig. 10-4. This arrangement places the console back from the window, the space between providing seating area for clients, producer, or other audience with minimum interference to the mixer. A larger observation window may be required to give these people close visual contact with the performers while keeping them out of the studio. If a really large window is decided upon, triple glass construction may be used to help offset the larger area. Excellent access to the rear of the console is provided simply by moving the observers' chairs. The raised platform provides ample space below the raised floor for cable runs; such a platform can provide needed low-frequency absorption by diaphragm action if needed.

Mixdown Rooms

As soon as the technical staff exceeds a complement of one, the need for a mixdown room will be apparent so that recording on one job and mixdown on another can proceed concurrently. Many decisions can prolong mixdown sessions, making a separate room for such postrecording work very valuable.

The mixdown room is as much a listening room as the control room itself—and similar specifications such as the size, shape, and acoustic treatment should prevail. Ideally, all mixdown and control rooms of a given complex or organization should match acoustically so that a given recording will sound the same to a given person no matter where it is played back. The mixdown function is fairly compatible with other production tasks such as tape editing and dubbing. The size of the mixdown room should possibly be larger than normal if much of this production work is contemplated. The number of rooms set aside for mixdown and other production jobs is entirely a matter of the size of the staff and the amount of business the company expects to have.

Reverberator and Reverberation Rooms

Full-size mechanical reverberation devices are usually considered too large and bulky to mount in the control room. Further, they can pick up local sounds, so it is generally better to house them in a small room tucked away somewhere in the studio complex. The EMT-140TS reverberation unit (Chapter 4) is $92 \times 51 \times 13$ in. and weighs approximately 380 lb. This plate-type reverberator can be located in a closet down the hall and connected to the control panel by shielded audio cable. The closet should be in a quiet location to avoid picking up ambient noises. The EMT-240 reverb foil is only $25 \times 12 \times 25$ in., weighs 148 lb, and is much less sensitive to pickup of local sounds. However, with the high monitoring levels common in control rooms today, it is also advised to locate this type of reverberator elsewhere and operate it remotely. There are many other reverberators utilizing springs, tapes, and digital techniques, but they are much smaller, less susceptible to sound pickup, and are typically rack mounted.

A reverberation room is quite another thing.[4] It is a chamber in which all six surfaces are highly reflective, preferably built after the fashion of Fig. 10-5 so that no two opposing surfaces are parallel to each other. The signal coming into the room from the loudspeaker is picked up on a microphone. The microphone should have a cardioid pattern, with the null directed toward the loudspeaker; if it is the omnidirectional type, an interposed barrier should be employed to reduce the direct pickup and to favor the reverberant sound. Reverberation times from 2 to 10 seconds can be achieved in chambers of reasonable size.

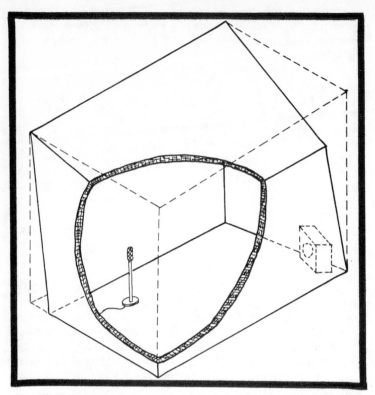

Fig. 10-5. Reverberation room having nonparallel surfaces.

Figure 10-6 shows how reverberation time depends on volume and type of surface. For simplicity, a cubicle room has been assumed and "eyeball" averages of absorption coefficients for various frequencies have been used for plaster and concrete surfaces. The graphs show that quite a reasonable reverberant chamber having reverberation times up to about 8 seconds can be made with concrete surfaces in the 1000 to 3000 cu ft volume range. The original construction determines the maximum reverberation time, and this can be readily lowered by introducing absorbing material into the chamber. This gives a rough size requirement for preliminary planning if such chambers are to be included in the plans of the studio complex. The canting angle of surfaces should be about 1 ft/10 ft. It is well to incorporate preferred room dimensions (Chapter 6) in determining the basic shape of the chamber.

The reverberation chamber is a three-dimensional system and has certain spatial advantages over the plate type (two-dimensional) and the wire type (one-dimensional),

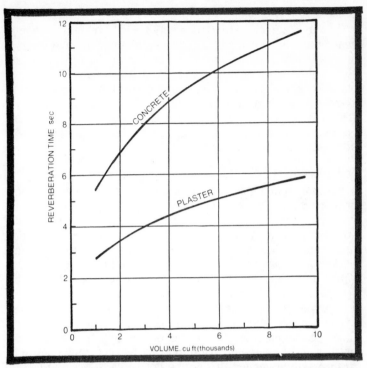

Fig. 10-6. Estimated reverberation time of reverb chamber as a function of volume and surface treatment.

according to some experts. However, chambers take up precious space and, where several are required (a common condition in modern multitrack studios), the cost may be prohibitive. Recent improvements in mechanical and electronic reverberators and the application of digital delay devices are steadily reducing the performance gap.

Mastering Room

Usually mastering work is sent out to specialists, but if the final mixed magnetic master is to be transferred to a master lacquer, a special room is required for this operation. This mastering room is really a small precision machine room which houses the magnetic playback and its ancillary electronic equipment as well as the disc lathe with its control and vacuum equipment. An area of 150 to 400 sq ft is required, depending upon the amount of equipment used. There are no special requirements as to ceiling height or acoustics unless a service of optimizing the tracks is offered, in which case good listening conditions are needed.

Electronic Workshop

To keep a modern recording activity in top condition, some space must be allocated for repair, maintenance, and testing of electronic equipment. The requirements for such an electronic workshop are quite nominal. It should have a carpeted bench (to avoid equipment scratches) and a shelf above the bench for test instruments. A minimum floor area of 100 sq ft should be set aside for this shop. This should be considered a noise producing area and should not be located adjacent to a studio or other low-noise area.

Other Rooms

A judgment will have to be made as to the necessity of one or more rehearsal rooms. If a studio is free, it can absorb any rehearsal time required. If solidly booked, separate rehearsal space is almost mandatory. As the rehearsal activity is not recorded, the acoustical requirements are relaxed with one exception. It is necessary to design it so that rehearsal sounds, which may be of very high level, do not interfere with nearby recording sessions or with neighbors.

Supporting services such as receptionist, secretaries, stenographers, and bookkeepers must be included in the plans of the studio complex. The office of the manager, as pointed out previously, can be very important as recording contracts are wooed and signed. It may be desirable to provide both a lounging space for the artists and a VIP room for the star's convenience and protection between sessions. And don't forget such vital facilities as the restrooms and the coffee dispenser.

SOME REAL-LIFE STUDIOS

To see how others have solved the problem of physically relating the various necessary functions of a recording operation, we shall look at several case histories in some detail. We shall see that many different solutions are possible—some elegant and expensive, others straightforward but still highly functional. In addition to general layout, we shall observe technical features of interest in both large institutional recording operations and more independent ones.

Japan Victor Company Studios

Recording music and selling records are a big business in Japan, and the great expansion in business which has been experienced has been accompanied by many new technological trends—including multitrack techniques. To keep abreast of such changes the Japan Victor Company saw, in 1969, the completion of a new five-story building of 39,000 sq

Fig. 10-7. Japan Victor Company recording studios.

ft devoted entirely to the firm's recording business (Fig. 10-7).[5] On the ground floor—the only floor with windows—are offices, dining room, and spacious lobby. Eight mastering and editing rooms, six reverberation chambers, and an artificial reverberator room are located on the second floor. Three recording studios, each with its own control room, and four rehearsal rooms are on the quieter third, fourth, and fifth floors.

Design Objectives. The objective upon which the design of the Victor studio complex is based are:

- Intermicrophone separation of 15 dB for multi-microphone recording.
- Mobile reflector baffles to simulate a live condition for more traditional recording with one or a few microphones.
- Reverberation time uniform throughout the audible spectrum. Particular attention was to be given to hold reverberation time to nominal values to frequencies as low as 50 Hz to avoid adverse effects of percussion and low-pitched strings on other instruments.
- Good diffusion of sound to make noncritical the placement of microphones, artists, and instruments.
- Combination of reverberation chambers, plates, and disc delay equipment to give flexible and convenient reverberation facilities.

219

Table 10-2. JVC Studio Specifications

Parameter	Large Studio	Medium Studio	Small Studio
Length, ft	53	33	18
Width, ft	75	43	20
Ceiling height, ft	30	20	10
Floor area, sq ft	3,900	1,400	360
Volume, cu ft	117,000	28,000	3,600
Reverberation time, sec	0.6	0.4	0.25
Avg absorption coefficient	0.45	0.45	0.37
Number of instrumental groups to be accommodated	12	6	2

- Studio background noise level to be held to a specific curve that is within 10 dB of the threshold of hearing.
- Mixer consoles to have 30 input channels and 16, 8, 4, and 2 output channels, selectable by operator.

Studio Size. The specifications for the three studios are outlined in Table 10-2:

To realize the 15 dB interchannel separation suggested by the studio mixers, it was necessary to determine experimentally the output of each instrument group, microphone placement, and directivity, and other such factors. From such data the dimensions of the studios were derived. Each instrument group was found to require a floor area of about 220 to 320 sq ft.

Acoustic Design. Conventional sound-absorbing structures as well as proprietary materials were found wanting in the search for uniform absorption down to 50 Hz. For this reason JVC engineers developed their own multilayer 3 ft thick structure, shown in Fig. 10-8. This was applied to 70% of the studio walls and ceilings, the remaining area of walls and ceilings being left reflective. The corrugated surfaces of both treated areas (Fig. 10-9) assures good diffusion and protection of the acoustical materials in the two larger studios. Further local diffusion is obtained by placement of mobile

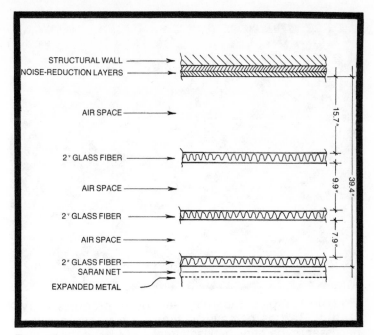

Fig. 10-8. Multilayer wall treatment of JVC studios to achieve near perfect absorption to very low frequencies.

Fig. 10-9. Corrugated surfaces of both treated and un-treated areas in the two larger JVC studios.

Fig. 10-10. Typical recording session in progress in JVC studio. Baffles of several types are used to obtain necessary separation between tracks.

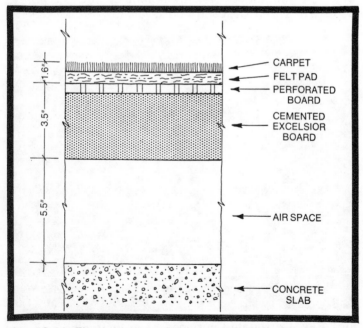

Fig. 10-11. Floor section of the small JVC studio designed to give high absorption over a wide frequency range.

Fig. 10-12. Mastering room for quadraphonic mixdown in the JVC studio complex.

polycylindrical baffle diffusers at the lower wall near the instruments, as shown in Fig. 10-10.

The smaller studio has an irregular, quadrilateral shape and a most interesting floor, shown in section in Fig. 10-11, which absorbs sound down to the lower frequencies. The carpet absorbs the higher frequency energy, and the perforated board and air space constitute a Helmholtz structure absorbent in the lower frequencies. The cemented Excelsior board broadens the absorption peak.

The extensive use of baffles to achieve the desired separation for a pop recording is shown in Fig. 10-10. Figure 10-12 shows a mastering room for quadraphonic mixdown. The final measured reverberation time versus frequency for all three studios is shown in Fig. 10-13.

This studio is an excellent example of lavish investment of time, engineering talent, and money to achieve the best results that modern technology allows.

Sound City

An excellent example of adapting an industrial building for studio use is Sound City studios of Van Nuys, California, in the suburbs of Los Angeles. This independent studio is owned and operated by Joe Gottfried and Tom Skeeter. The building is in the center of a cluster of modest industrial buildings, but it stands alone and is shielded somewhat from street and freeway noise by the other buildings. The entire operation has

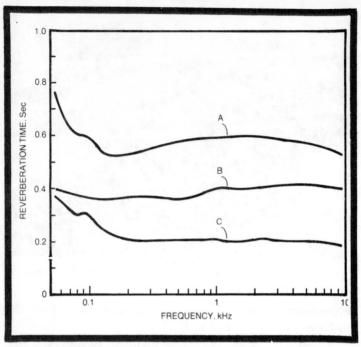

Fig. 10-13. Reverberation time of three JVC studios as a function of frequency.

an honest, businesslike air about it, giving the impression of competence without fancy frills.

The 60 by 115 ft building is depicted by floor plan in Fig. 10-14. It is built around the very common two-studio plan: one large and one small studio, each with its own control room. The entrances to all studios and control rooms are through *sound lock* corridors, with the exception of the rear door of control room A, which opens into a low activity storage area and hall. The two studios are isolated from each other by halls, other rooms, and considerable distance. A nice feature for those tedious recording and mixdown sessions is a lounge area, equipped with a billiard table, comfortable seats, and food dispensing machines. A double sound lock arrangement connects the lounge (which might be a bit noisy) to studio A and its control room.

Studio A is about 40 ft square, except that the east wall is splayed 1 ft/8 ft, and the ceiling is about 14 ft high. Rough lath, part of a low-frequency absorbing structure, provides a textured surface for the east wall, equipped with drapes that can be extended or retracted at will. The north wall is

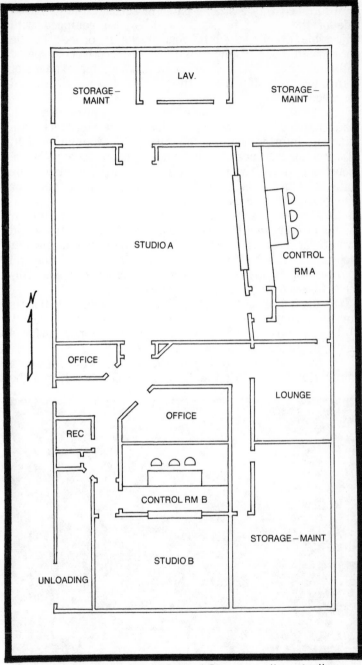

Fig. 10-14. Floor plan of Sound City recording studios.

hard-surfaced with a sizable patch of acoustic tile on either side of the door. The south wall is also equipped with adjustable drapes, used to control flutter echoes in the north/south mode. The vinyl-tile floor is selectively carpeted with large rugs. The ceiling has a considerable expanse of perforated acoustic tile with panels of soft fiberboard suspended below.

In the studio A control room the console is raised on a 12 in. platform which provides space for a few observers with minimum interference with the mixer. Two biamplified JBL 2230 monitor loudspeakers on the left and two on the right are mounted in a soffit which aims them toward the mixer. The soffit continues around the entire room, providing space for rear loudspeakers at a later time.

One recorder is in an alcove to the right rear of the mixer, but the primary multitrack machines—an Ampex 24-track and two Studer 2-tracks—are to the left of the mixer in another alcove. All in all, the arrangement has given satisfaction and the operators have found it to be convenient and efficient. This facility is used for mixdown work as well as recording.

Studio B is about 19 by 28 ft, with a 14 ft ceiling. The floor is covered with vinyl tile; wall treatment includes low-frequency absorbers and fabric-covered glass fiber. Inclined surfaces contribute to the diffusion of sound in the studio.

In the studio B control room, the console is mounted on a raised platform, with an audience area between it and the

Fig. 10-15. Control room of studio B, Sound City.

observation window, similar to the arrangement in the studio A control room. The front and rear loudspeakers are built in. Three overhead reflectors (Fig. 10-15) dominate the area above the console.

Both studios are equipped with Neve consoles (24/16 in A, and 24/8 in B); the recording and ancillary equipment includes Ampex machines of 8, 4, and 2 tracks and two Studer 2-track machines for mastering. The ceiling of the control room is cork, the floor is heavily carpeted, and the fabric-covered areas between the vertical 4 × 4s on the side walls are highly absorbent. There are 24 lines running between the two control rooms.

Sound City's leased recording complex represents an investment of about $700,000, some 25% of which went into adapting the building, studio construction, and treatment. To show how such an activity can soak up money in details, consider the cost of the facility's more than 80 microphones!

Paramount Recording Studios

The only thing the Paramount Recording Studio organization holds in common with the motion picture studio of the same name is that they are both in Hollywood. This recording studio is an independent business dominated by Brian Bruderlin, who is owner, studio manager, and chief engineer. Paramount purchased an existing building and adapted it to specific needs.

The building (Fig. 10-16), rear parking lot, and driveway are contained in a plot 58 by 135 ft, no larger than an average

Fig. 10-16. Paramount Recording Studios, Hollywood, California.

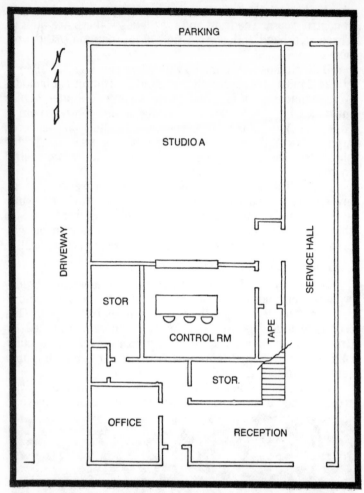

Fig. 10-17. Ground floor plan of Paramount Recording Studios.

residential lot. The building itself is 48 by 78 ft with the attractive facade facing busy Santa Monica Boulevard. Paneled walls, drapes, and carpets give no hint that the building itself dates back to 1946. The main entrance is on one side of a lounging area and a generous hall running the full length of the building is on the other. This hall, as shown in the floor plan of Fig. 10-17, opens to the boulevard in front and the parking area in the rear, and provides easy access to all the studios for personnel and bulky instruments. Figure 10-18 shows the floor plan of the facilities on the second floor.

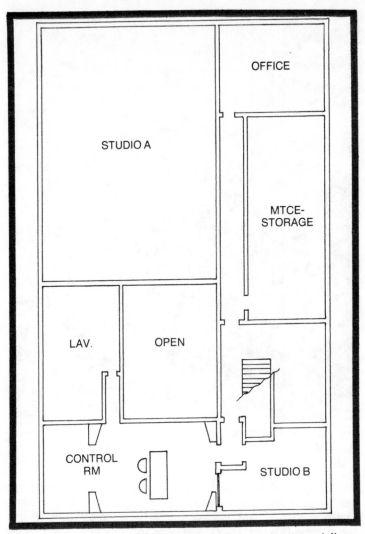

Fig. 10-18. Paramount Recording Studios' second-floor facilities.

Passing from the ground-floor corridor through a generous sound lock which serves studio, control room, and tape storage room, we enter studio A. It is 35 by 45 ft in size with a ceiling height of about 22 ft. The ceiling follows the dome-shaped contour of the roof as seen in Fig. 10-19 which shows the northeast corner of the studio. The isolation booth door is plainly marked for those late-night sessions when vision is a bit hazy.

Fig. 10-19. Northeast corner of studio A, Paramount Recording Studios, showing isolation booth and roof treatment.

The control room of studio A is dominated by an imposing custom Spectra-Sonics 24-in/16-out console on a raised platform with a small audience area between the console and the observation window. A Stephens 16-track recorder and an Ampex 2-track machine carry the major recording load of this studio. The loudspeaker complement is four built-in Altec 604Es and two outboard JBL 4310s.

Studio B on the upper level is reached by a stairway off the main corridor. It is small and of irregular shape, approximately 15 by 15 ft, and is used primarily for overdubbing. Figure 10-20 is a view of the upstairs studio through a novel leaded-glass observation window in the control room. The blurred foreground image is a Spectra-Sonics 20 in/8 out console. Beyond the rear loudspeakers is an ample

Fig. 10-20. View looking west in control room of Paramount's studio B showing the Stephens 16-track, Scully 8 and 4-track, and Ampex 2-track recorders. An observers' area is in the far end of the room beyond the loudspeakers.

audience area. The recording machines include a Stephens 16-track, Scully 8-track and 4-track machines, and an Ampex 2-track system equipped with Dolby for mastering. Loudspeakers include four built-in Altec 604s and 2 JBL 4310s.

Also on the upper level is a maintenance shop and a second office.

Paramount Recording Studios, established in 1966, can almost be considered a pioneer in this fast-moving field. Brian Bruderlin is constantly in the process of upgrading facilities and improving service. Automated mixdown is presently being seriously considered. The total investment amounts to about $350,000.

REFERENCES

1. Rettinger, Michael. *Acoustic Design and Noise Control.* Chemical Publishing Co., Inc., New York 1973 (pp 409–410).

2. Gilford, C. *The Acoustic Design of Talk Studios and Listening Rooms.* Proc. Inst. of IEEE, Vol. 106, Part B, No. 27, May 1959 (p253).

3. Billboard 1973 and 1974 editions of *International Directory of Recording Studios.*

4. Rettinger, Michael. *Reverberation Chambers.* Jour. Audio Engr. Soc., Vol. 20, No. 9, November 1972 (pp 734–737).

5. Shiraishi, Y., K. Okumura, and M. Fujimoto. *Innovations in Studio Design and Recording in the Victor Record Studios,* Jour. Audio Engr. Soc., Vol. 19, No. 5, May 1971 (pp 405–409).

11 Multitrack Studio Acoustics

The acoustics of the space and the type of program produced in the space have irrevocably been tied together from antiquity. Classical Greek theaters have been found to have exceptionally good acoustics for speech intelligibility, solo or unison singing, chanting, and for solo musical instruments.[1] The design of these structures produces a profusion of early reflected sound, which reaches the ears of the listeners within 50 msec of the direct sound. The reflecting surfaces suspended over the heads of the audience in some of the latest and best designed auditoriums are, in essence, a return to this hoary principle. The amazing fact is that the ancient theater at Epidaurus, seating 14,000 people, provided good intelligibility of speech at a distance of 200 ft with no sound reinforcement system other than that provided by its geometry!

Such a Greek theater would be poor for orchestral music, as natural reverberation is almost totally absent. A fascinating sidelight on this is that Vitruvius described a method by which a degree of artificial reverberation was introduced in Greek theaters by deploying bronze pot resonators around the amphitheater. These resonators were tuned to different frequencies and were distributed in definite patterns.[2] Did the acoustics of outdoor Greek theaters shape the programing? Or did they build the theaters with acoustics to fit their programing? The evidence favors the former, but of one thing we can be certain: there is a close relationship between the type of program and the acoustics of the space.

A composer adapts himself to circumstances and naturally composes music to sound "right" in the space in which it will be played. Early church music was written to fit the large and highly reverberant cathedrals, operas to sound right in the opera houses, and chamber music to fit the drawing room. Bach's more ponderous organ music would be as ill-fitted to the drawing room as chamber music would be to the cathedral. The rapid and delicate fingering of strings would be drowned within a slurring sea of reverberation in a cathedral having a reverberation time of 4 to 6 seconds. The wood-paneled walls of the old drawing rooms were excellent low-frequency absorbers, nicely offsetting the high-frequency absorption of drapes, curtains, and rugs; and out of this setting grew an intimate and charming form of music that is closely tied to the acoustics of the space in which it is performed.

Today we have a new element injected in recorded sound. Music played in the home from discs or tapes or music reproduced via radio or television is the only music heard by most people. In America (perhaps less so in Europe) relatively few people hear live musical performances and more and more musical tastes and standards are shaped by what is on the tape or disc. What was done to the music along the way? What effect did studio acoustics have? What equalization did the control board operator inject? What special effects were introduced in mixdown? How did the reproducing elements, amplifiers, loudspeakers, and living room acoustics affect the music? In the old days, how it sounded in the music hall was the criterion. What is the criterion today?

From the classical perspective, things musical could be considered to be in a chaotic condition as every man does that which is right in his own eyes (ears). On the other hand, music is reaching the masses through the recording media and millions are enjoying music which was once the province of only the privileged few. Recording is now a great industry—this very book, in fact, is designed as a guide for those in or wanting to get into the recording field. This particular chapter explores the issue of what kind of acoustics we should build into our studio. First, we must decide what we will be recording, for that determines the acoustics we'll require.

To reproduce the concert hall condition, a single microphone for monophonic or a simple crossed pair of cardioid microphones for stereophonic recording is usually the choice of critical judges accustomed to hearing live concerts. It is possible that acoustic deficiencies of the hall might require a few more microphones, especially for monophonic, to give good overall concert hall balance. The point here, however, is that the balance is fitted to the hall, the audience, and the microphone position. The music is premixed as it falls on the microphone. For such recording, a good concert hall or studio is required which places it outside our subject of recording studios, although many operators of studios do offer a mobile recording service.

There is no way to achieve exactly this same balance in a much smaller studio, but it can be approached and possibly improved upon. The orchestra can be recorded in a dead studio by breaking it down into groups, each with its own track on the recorder. Later the various tracks can be combined at appropriate levels in 2- or 4-channel stereo form with reverberation added to restore the large hall effect. Note that how the tracks are mixed is now determined by the experience, orientation, and standards of the one doing the mixdown. If he tries only to make a concert hall sound, he will be sacrificing much of the inherent flexibility of the multitrack system. He now has the freedom of adapting the orchestra sounds to the living room, of moving the emphasis around to parts of the orchestra engaged in the most interesting activity at the moment.He accomplishes with his controls what an alert, music minded auditor listening in the concert hall involuntarily does through the marvel of his psychoacoustic senses, shifting his attention from section to section, or player to player, to catch features of maximum interest while rejecting, to a degree, the other features.

The fashions and fads of popular music are continually shifting. During the past decade considerable experience has been gained in recording rock bands commonly using instruments with electrical amplification. The main problem in such recording is to achieve adequate separation. Low room reverberation time, physical separation, baffles, directional microphones, and certain electronic devices are commonly employed to yield the required separation.

STUDIO REVERBERATION TIME

In the preceding chapter we considered the sizes of hundreds of U.S. recording studios; unfortunately, statistics on the reverberation characteristics are much harder to come by. Rarely are acoustical designs of recording studios published. In Great Britain the acoustical characteristics of hundreds of British Broadcasting Corporation studios are determined largely by staff engineers, and operating such a large number of studios has stimulated significant investigations of acoustical problems by the BBC research teams as well. Their experience can be a partial guide for us.

In Fig. 11-1 the shaded area A is a rough indication of the range of the reverberation times of enclosures for various

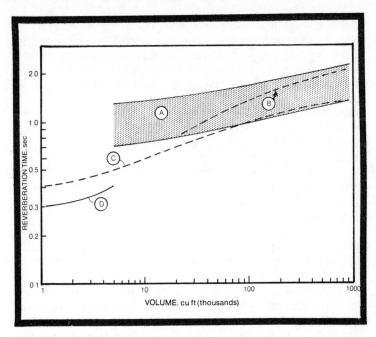

Fig. 11-1. Reverberation time of traditional enclosures as a function of room size. Area A: upper edge suitable for organ music; near center for smaller enclosures suitable for chamber music, large enclosures for symphony and opera; lower edge, traditional broadcasting and sound recording studios. Curve B: BBC studios for one-point pickup. Curve C: NBC[5] optimum (1936). Curve D: BBC talk studios.

general purposes as a function of volume. The larger enclosures along the upper portion of this shaded area are suitable for organ music. Smaller enclosures slightly above the center line of the shaded area would be suitable for chamber music and the larger ones for symphony orchestras and opera. Motion picture theaters fall somewhat below the center line. Close to the lower edge are the traditional broadcasting and sound recording studios.

Curve B is considered by the BBC to be the optimum reverberation time (in the frequency range of 500–2000 Hz) for music studios suitable for broadcasting and recording of the one-poinnt pickup type.[3,4] Curve C is the NBC optimum (1936).[5] Curve D is the BBC curve for talk studios, generally of the 1000–5000 cu ft range. Figure 11-1 helps us to fix in mind the range of reverberation times for the more traditional types of studios.

Dead studios have already been mentioned as a requirement for adequate separation in multitrack recording. If we want to achieve maximum separation, how dead is it possible to make a studio? For one thing, we know that maximum absorption requires a wideband absorbing material, absorbing close to 100% of the sound falling on it throughout the audible band. Unfortunately, there is no such perfect absorber. Sabine's original concept for a perfect absorber was an open window, but open windows seem particularly inappropriate for our studio!

In the more traditional studios having longer reverberation times, it is common practice to balance materials having good high-frequency absorption (porous type) with others having good low-frequency absorption (resonator type), distributing these among the various surfaces. In a superdead studio such profligate use of limited area is unthinkable as we need to apply wideband absorbers to every available square foot.

Going now to Fig. 11-2, we see the range of reverberation times for various types of traditional studios, area A, repeated for comparison. Area B is bounded by two computed, hypothetical cases, one for studios of various sizes having 70% absorbers (coefficient of absorption, 0.7) on five of the six surfaces, and one for 90% absorbers on five surfaces. For ease of calculation a rectangular studio having a ceiling height half the width and a length twice the width is assumed—although this is by no means a recommendation for such proportions.

The broken line (C) running through area B is considered by the BBC to define the lowest practical reverberation times for television studios; line D represents their idea of the optimum for television studios. In Chapter 10 we studied in some detail the design of the Japan Victor Company studios; the measured reverberation times of these three studios (in the 500—2000 Hz region) are shown in Fig. 11-2 as points 1, 2, and 3. They fall close to or just under the BBC optimum for television studios. It will be recalled that JVC engineers were reluctant to abandon one-point microphone recording in spite of the popularity of multitrack techniques today, and we may consider their design as something of a compromise. In fact, their specifications for reverberation are closer to the BBC optimum for television studios than the actual measured values plotted in Fig. 11-2.

Point 4 in Fig. 11-2 represents common practice in the acoustical treatment of motion picture sound stages.[6] Point 5 is the range for the new A&M Records studio in Hollywood.[7] Point 6 is for the RCA studio in Nashville.[8,9] Our general

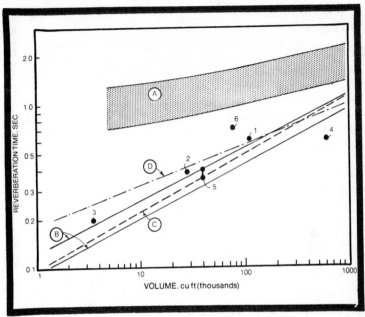

Fig. 11-2. Reverberation time of modern studios as a function of room size. Area A: Same as Fig. 11-1 for comparison. Area B: Computed for studios moderately dead to exceptionally dead. Curve C: BBC lowest practical for TV studios.Curve D: BBC optimum for TV studios. Measured points: (1) Japan Victor Co. large studio; (2) JVC intermediate studio; (3) JVC small studio; (4) Motion picture sound stage[6]; (5) A&M Records studio[7]; (6) RCA Records, Nashville[8];

conclusion is that our computed area B in Fig. 11-2 gives a fair representation of recording studios devoted to popular music.

MOTION PICTURE SOUND STAGE

Until very recently, practically all studio design energies have been directed toward the traditional blending type of microphone pickup which has dominated the radio, television, and recording industries. The trend is now toward multitrack recording on a separation basis for disc and tape releases and increasingly for the more important television productions as well. We shall concentrate on the studio acoustics best suited to the multitrack approach because it is universally used in the pop field and increasingly so in more serious music as well.

Very low reverberation is required to achieve the desired separation between channels. The 600,000 cu ft motion picture

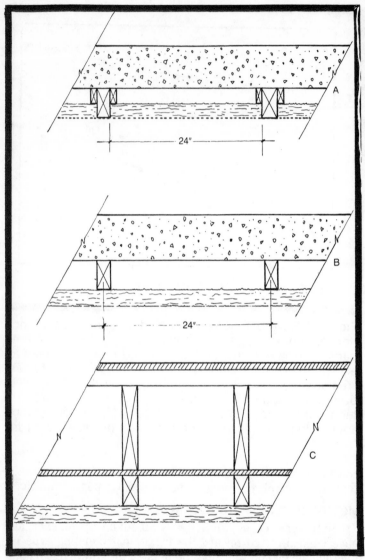

Fig. 11-3. Typical motion picture sound stage acoustic treatment: (A) lower walls with heavy wire screen protection, (B) upper walls, and (C) roof/ceiling.

sound stage represented by 4 in Fig. 11-2 achieves a reverberation time of about 600 msec (0.6 sec) by application of 2 in. glass fiber board (minimum density 4 lb cu ft) to all surfaces but the floor. The detail of Fig. 11-3A shows how

2-by-4s and 2 in. furring strips support the 24 in. sheets of glass fiber board with an air space of 2 in. behind the board. The glass fiber board may be covered with a glass fabric having the proper fire characteristics and which is transparent to sound.

The structure of Fig. 11-3A is especially suited for the lower walls so that a heavy wirescreen can be nailed to the 2-by-4s to protect the glass fiber from damage in normal stage activity. The structure of Fig. 11-3B provides a 4 in. air space by mounting the glass fiber board to the tops of the 2-by4s. This gives better low-frequency absorption and is quite practical on the higher reaches of the wall where the protection of the wire screen is not needed. Figures 11-3C describes a typical treatment of the roof. Sheets of fiber board are nailed to the bottom of the roof rafters below which the 2-by-4s supporting the absorbing material are secured. Sometimes a 2 in. layer of concrete is poured on the roof for added protection against outside noise.

THE EYE vs THE EAR

This much detail is given for the motion picture sound stage because the dead acoustical characteristics of a sound stage are similar to what is needed in the multitrack recording studio. It won't look like much, but it will work. The truth is, however, that the esthetic sensibilities of the artists and customers must be considered, and there is little about the appearance of the inside of a sound stage that would urge them to greater artistic heights. There are various ways that unattractive but acoustically good surfaces can be covered, and a good architect can dress up such surfaces to give a visually pleasing effect. The important thing to remember is that the covering must be transparent to sound and it must not introduce reflecting surfaces which could be troublesome.

Figure 10-8 (preceding chapter) shows how JVC covered walls with Saran net for dust protection and with expanded metal for visual treatment and mechanical protection. The overall effect in the JVC studio is decidedly modern and pleasing.

Another approach at camouflage of dismal acoustical surfaces is covering them with wooden fins or slats backed with burlap or other fabric.[10] It is important to keep the fins shallow (1 or 2 in.) and the spaces as large as the visual effect will allow to minimize their action as Helmholtz resonators. The other approach is to make the fins into Helmholtz resonators and incorporate them into the acoustic design of the room—this requires carefully adjusted dimension combinations of slit width and depth and cavity depth.[11,12] Just do not erect a big area of slats and slits without due attention to a possible resonator effect! These resonators can be made to peak in the low frequencies, which may be desirable to compensate for the normal absorbing materials.

EXCURSUS: ACOUSTICAL PRINCIPLES

The search for very short reverberation time in the multitrack recording studio resolves itself into a search for highly absorbent materials and structures which meet the fourfold test of *absorption, cost, appearance,* and *strength.* The motion picture sound stage approach is effective and cheap, but not very attractive. Before we proceed with specific solutions, it is well to review a few acoustical facts of life and, at the same time, look at some available materials.

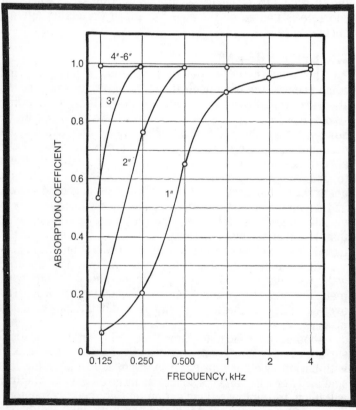

Fig. 11-4. Effect of thickness on sound absorption coefficient of Owens-Corning Fiberglas Type 703, 3 lb/cu ft density.

First, a rather self-evident fact: In general, the thicker the absorbing material, the better the absorption at the low-frequency end of the audible spectrum. This makes sense. A glass fiber board 1 in. thick has an acoustic thickness at 10 kHz of approximately one wavelength (1λ). At 100 Hz, however, the acoustical thickness is only about 0.01λ. This point is verified by actual measurements made in the Owens-Corning acoustical laboratory[13] on Type 703 insulating board (density of 3 lb/cu ft), shown in Fig. 11-4. The absorption coefficient of the 1 in. board falls off rapidly at frequencies below about 1 kHz. Doubling the thickness results in good absorption down to 300—400 Hz. A thickness of 3 in. yields a coefficient of 0.9 or better to a frequency below 200 Hz. At the lowest measuring frequency, 125 Hz, material of both 4 and 6 in. thickness absorbs 99% of the sound energy falling on it. We can be sure that the 6 in. board absorbs to a somewhat lower frequency than the 4 in. board. As far as the 125—4000 Hz band is concerned, we have something that approaches Sabine's perfect absorber—the open window.

Absorption measurements below 125 Hz become increasingly difficult, yet recording equipment and our ears go down a couple of octaves below this frequency. It is unfortunate that absorption information is not often available for lower frequencies of interest. About the only course of action open to us is to make a calculation here and an educated guess there and verify the result by actual measurements of reverberation after the acoustic treatment is installed. Any appreciable rise of reverberation time in the bass frequencies must be avoided. The smaller the studio, the more important is the *flat reverberation time vs frequency* characteristic.

In Fig. 11-3 we note that in this motion picture sound stage there is an air space between the absorbing material and the structural surface. Such an air space increases the absorbing efficiency of the material, and air space increases the absorbing efficiency of the material and air space is usually cheaper than, say, doubling the thickness of absorbing material. Figure 11-5 illustrates this point with measurements on Owens-Corning glass cloth faced Fiberglas boards.[13] As the air space is increased from 1 to 5 in., we see the steady improvement in low-frequency absorption as 90% absorption is moved from about 650 Hz almost down to 300Hz—or about one octave. There is a limit to how much air space we can use advantageously behind a layer of a certain thickness, however, because of dips which develop in the absorption graph at other frequencies.

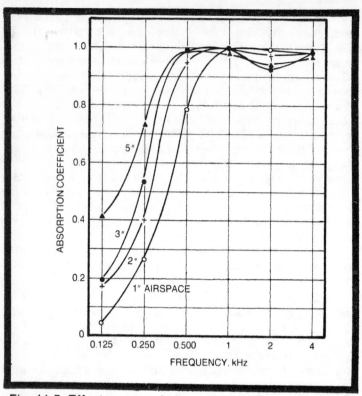

Fig. 11-5. Effect on sound absorption coefficient of the air space behind Owens-Corning Fiberglas linear glass cloth board of 1 in. thickness.

The board of Fig. 11-5 and the thicker Type 703 Fiberglas material of Fig. 11-4 are combined in Fig. 11-6. Graph A, repeated from Fig. 11-5, shows the 1 in. glass/cloth-faced board and a 5 in. air space behind it. In graph B the same board is backed by 2 in. of Type 703 board. Graphs C and D show the effects of increasing the Type 703 backing to 4 in. and 5 in., respectively. There are no air spaces in the B, C, and D examples, as the space is filled with Type 703 board. With 5 in. of backing absorbent, almost perfect absorption results from 125 Hz upward. These results can be approached by the cheaper expedient of using thinner absorbent and more air space, although thicker wall treatment encroaches upon studio space.

Figure 11-6 suggests an approach to a combination which gives good appearance, good absorption (down to 125 Hz at least), good mechanical strength, and resistance to wear. The

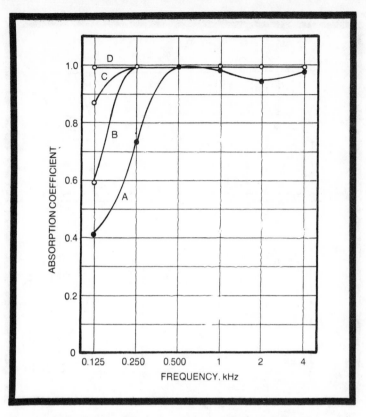

Fig. 11-6. Sound absorption characteristics of combination structures: (A Owens-Corning unpainted linear glass cloth board (1 in.) plus 5 in. air space; (B) the same with 2 in. Type 703 Fiberglas board; (C) with 4 in. Type 703 Fiberglas board; (D) with 5 in. Type 703 Fiberglas board.

cost is relatively modest when we consider the ease of installation. One form of mounting in 4 by 8 ft panels is sketched in Fig. 11-7. Battens may be nailed or screwed to the 1 by 1 in. mounting strips nailed to the 2 by 6 in. members. The glass cloth-faced board has a very attractive surface, and if molding of the right shape and finish is used as battens, a very pleasing installation results.

There is a problem with respect to low-frequency absorption obtained through the use of air spaces. Resonance modes parallel to the wall may be set up within the air space, reducing the absorption efficiency—especially at the lower frequencies. Such modes may be controlled by breaking up the

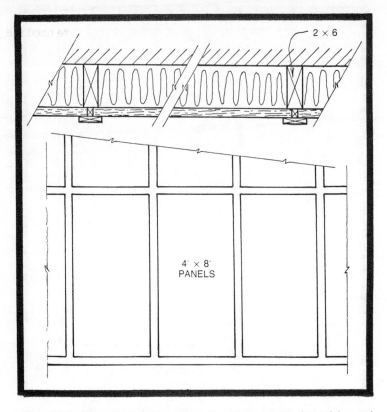

Fig. 11-7. Possible mounting of a glass cloth-faced board with backing to provide attractive studio treatment.

air space by egg crate baffling of thin wood or corrugated paper.

Another more straightforward approach is filling the air space with inexpensive rock wool or other common building insulation material. The latter would tend to extend the good absorption region of a given structure to even lower frequencies (Fig. 11-4). In fact, the new A&M Records studio in Hollywood uses 1 in. rigid glass fiber board (Owens-Corning Type 705) spaced from the wall as much as 12 in. with glass fiber blanket in the cavity.[7]

COMPOSITE STRUCTURES

To get the low reverberation time in the studio (which good separation demands), a high percentage of the area must be treated. If we start with the goal of maximum intertrack separation, we may no longer have the freedom of placing general treatment on some surfaces and low-frequency absorbers on other surfaces, no matter how much we need the low-frequency units in the two octaves below 125 Hz. In the smaller JVC studio, even the floor is elaborately treated. Not investing in complex floor structure leaves that much less freedom to leave walls and ceiling areas untreated.

It is possible to cover one type of absorber with an absorber of another type. In Fig. 11-8 a 6 in. air space is placed behind the arrangement of Fig. 11-7 with a thin panel such as 3/16 in. plywood between. The panel diaphragm and the air space resonate in the general region of 50 Hz. This diaphragm/air space arrangement alone would absorb very well at the frequency at which the structure resonates, but the absorption would be over a very narrow frequency range with no absorbent material in the cavity. Placing the structure of Fig. 11-7 over this panel would tend to broaden the absorption peak. One might think that the diaphragm would be shielded from the sound in the room by 6 in. of absorbent material; but remember that even 6 in. of glass fiber is relatively transparent to sound near 50 Hz.

TAPERED ACOUSTICS

The standard approach to the acoustic treatment of studios has been to distribute the various absorbing materials over the surfaces of the room to contribute to the diffuseness of sound. This approach considers the room as an entity and is an attempt to provide uniform sound conditions in all parts of the room.

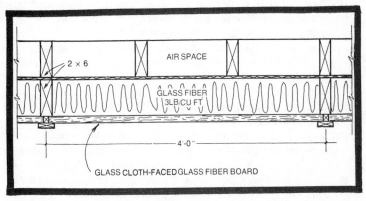

Fig. 11-8. A possible composite structure in which the material of Fig. 11-7 is placed over a low-frequency absorber consisting of a 3/16 in. plywood diaphram and an air space.

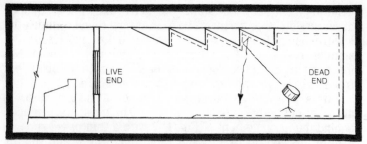

Fig. 11-9. Tapered studio acoustics provide a dead end and a live end. A serrated ceiling minimizes forward reflections.

Radio broadcast people have been chipping away at this concept for many years in studios built especially for radio drama. The ear and the imaginative consciousness of the listener provide the only channel in radio for establishing settings and characters, developing the story line, and communicating mood in dramatic presentations. For this reason, the effect of the acoustics of the microphone environment on voices and sound effects takes on great importance. During the course of a radio drama it may be necessary to simulate a voice outdoors, down a well, in a subterranean chamber, or to give other acoustic impressions.

The specialized radio drama studio has been developed, especially in Europe,[14] to provide the needed variations in acoustic perspective, and usually they have followed the live-end/dead-end principle. The acoustics of the live end of a long studio are bright and reverberant, the other end dead and absorbent. Curtains of heavy canvas, often in pairs separated by several feet, are available to provide separation between the live, medium, and dead areas of the studio for special applications. Often the long studio with its acoustics is supplemented by annex rooms having highly reverberant or very dead acoustics.

The studio with tapered acoustics is having some application in the pop music recording field. For example, Record Plant studios in New York and Los Angeles use this principle.[15] The acoustics are bright near the control room window and get progressively deader toward the far end, as in Fig. 11-9. A sawtooth-shaped absorbent ceiling has teeth pointing toward the dead end. In use, the rhythm section—which needs a "tight" sound—is located at the dead end of the room, and the strings are placed close to the control room.

Such a tapered studio would seem to be best adapted to a relatively fixed recording situation. This arrangement may be

246

poorly suited for groups of other sizes, other combinations of artists, and for other forms of music. Flexibility is the key word in multitrack recording, and some of this flexibility is surely sacrificed in going toward tapered acoustics, however attractive for the fixed format specialization.

REFERENCES

1. Shankland, Robert. *Acoustics of Greek Theatres.* Physics Today, Vol. 26, No.10, October 1973 (pp 30—35).

2. Brüel, Per. *Sound Insulation and Room Acoustics.* Chapman and Hall. London, 1951 (pp 216—219).

3. Burd, A., C. Gilford, and N. Spring. *Data For The Acoustic Design of Studios,* BBC Engineering Division Monograph 64. November 1966 (pp 16, 18, 19).

4. Brown, Sandy. *Acoustic Design of Broadcasting Studios.* Jour. Sound & Vibr., Vol. 1, No. 3, 1964 (pp 239—257).

5. Morris, R. and G. Nixon. *Broadcast Studio Design.* RCA Review, 1936.

6. Bloomberg, D. and M. Rettinger. *Modern Sound Stage Construction.* Jour. SMPTE, January 1966 (pp 16, 18, 19).

7. Christoff, Jerry. *Acoustical Design of a New Studio For A&M Records.* Paper M3, presented at the 48th convention of the Audio Engr. Soc., Los Angeles, May 1974.

8. Volkmann, John. *Acoustic Requirements of Stereo Recording Studios.* Jour. Audio Engr. Soc., Vol. 14, No. 4, October 1966 (pp 324—327).

9. Brown, Sandy. *Recording Studios For Popular Music.* Paper G36, 5th Intl. Congress on Acoustics. Liege. September 1965.

10. Staff. **dB** *Visits Ultrasonic Recording Studios.* **db** The Sound Engineering Magazine, Vol. 5, No. 11, November 1971 (pp 25—27).

11. Rettinger, Michael. *Low-Frequency Sound Absorbers.* **db.** The Sound Engineering Magazine, Vol. 4, No. 4, April 1970 (pp 44—46).

12. Smits, J. and C. Kosten. *Sound Absorption by Slit Resonators.* Acustica. Vol. 1, 1951 (pp 114—122).

13. *Staff. Wall Treatments,* Publ. 5-AC-4250C, available from Owens Corning Fiberglas Corp., Architectural Products Div., Fiberglas Tower, Toledo, Ohio 43659.

14. Gilford, Christopher. *Acoustics For Radio and Television Studios.* IEE Monograph Series 11, Peter Peregrinus, Ltd., London, 1972.

15. Hope, Adrian. *Record Plant and Westlake Audio.* Studio Sound, Vol. 14, No. 8, August 1972 (pp 29—30).

16. Hidley, Tom. *Acoustics in Studio Design.* Recording Engineer/Producer, Vol. 5, No. 3, June 1974 (pp 29—32, 37).

17. Harrison, Dave. *Acoustic Design: The Myth of the Magical Studio.* Recording Engineer/Producer, Vol. 5, No. 4, August 1974 (pp 49, 51—55).

12 Constructing the Studio Complex

Constructing a practical recording studio always involves more than the studio alone. A complex of rooms and areas is required. Operation of the studio requires offices to obtain jobs and to care for them after they are landed in such things as day-to-day executive direction, accounting, and other services. There must be space for technical support such as tape editing and mixdown and maintenance of facilities. The word *complex* is an appropriate descriptive noun for the assemblage of these related areas; the adjective form of the same word is also applicable in describing not only the intricate interrelationships of functions, but the constructional features required to orchestrate the performance of the whole. Although somewhat intricate and complex, these required constructional features really have the sort of logic behind them that any serious reader can readily understand. This chapter presents the reasons behind the essential features and discusses the importance of each as a guide for anyone planning to establish a recording studio and for the architect and building contractor involved. Knowledge of such things is also of value to those operating a studio; poor maintenance can, for example, destroy the acoustical integrity of a studio or control room.

ACOUSTICS AND CONSTRUCTION

We must look at the acoustical aspects of a recording studio complex from three points of view: (1) the penetration of unwanted outside noise into the recording studio, (2) the sounds generated within the studio which might disturb neighbors, and (3) the acoustical treatment of the studio itself and its effect on the sounds recorded in it. Some constructional details involve both (1) and (2); others do not. For example, a good heavy wall will protect the studio from traffic noise and shield neighbors from the sounds generated within the studio. On the other hand, noise from air-conditioning equipment might be a serious problem in the studio even though it will not bother the neighbors. Point (3) is treated in Chapter 11.

At this point it is well to recall the tenuous and intangible nature of all things acoustical. It is easy, especially for the layman, to entertain the thought that acoustical things are important in the recording business without having a genuine conviction of it. While giving lip service to such things, the true nature of the individual's priorities come to light when a bit of expense is involved beyond that necessary to keep the building from collapsing. Acoustics are intangible, but they are every bit as real as other intangibles such as air, capital gains, or honesty—and they accordingly must be given due consideration if the recording studio is going to yield acceptable results.

CONTROL OF IMPACT AND AIRBORNE NOISE

Let us first consider walls (exterior walls, inside partitions, ceilings, or roofs) simply as generalized barriers to the transmission of sound from one side of the barrier to the other. Figure 12-1 illustrates how one activity can interfere with another, a problem that is increasing with congestion. In Fig. 12-2 the sound level on side 1 of such a barrier is higher than the level on side 2 because of the dissipation of sound energy as it passes through the barrier. If the barrier is a wall between studio A and studio B, we may be interested in determining how a high-level rock band in A might interfere with a narration recording job going on at the same time in B. (We should note in passing that, as far as we are concerned in this book, a given barrier offers the same transmission loss to sound going from side 1 to side 2 as in the reverse direction.)

How can we determine the noise level in a given area of operation? What constitutes an acceptable background noise level inside a recording studio? Even if we knew the answers to these questions, how could this information guide us in specifying the type of walls and doors our studio should have? The remainder of this chapter is devoted to the basic principles involved and will, we hope, make it possible to arrive at definite answers to the specific problems one faces in establishing a new studio and operating an existing one.

Figure 12-3 clarifies the nature of our problem. We must have some knowledge of the general noise levels of the area in

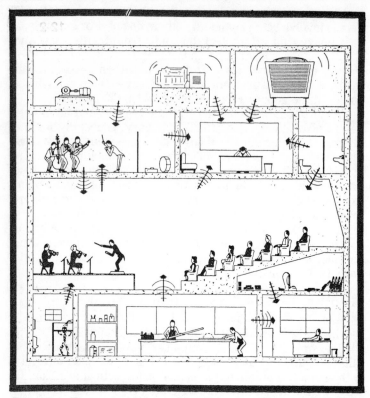

Fig. 12-1. The noise from one activity can be a very serious problem in its effects on other activities. (Consolidated Kinetics Corp.)

which the studio will be located, either by measurements or by judicious guessing and inference. Is the studio to be next to a busy interstate freeway? A well-traveled street? Industrial area? Or will it be in the wilds of Colorado like the Caribou Ranch studio? The noise from switching freight cars on a nearby track, or diesel trucks on a highway are easy to comprehend as enemies of a recording operation, but an occasional cow mooing, dog barking, automobile sound, or rooster crowing (the latter occurring primarily during mixdown sessions!), are very real hazards in the quietest locations. However, in the quieter locations thinner walls may suffice because of the generally lower noise environment.

NOISE SPECTRA

Returning to Fig. 12-3, we note that noise energy varies with frequency. This implies some method of breaking the

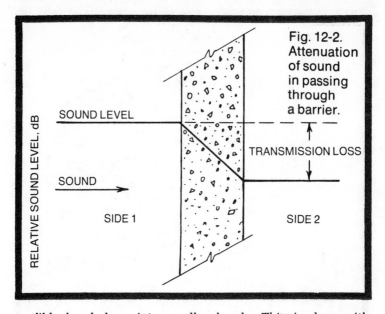

Fig. 12-2. Attenuation of sound in passing through a barrier.

audible band down into smaller bands. This is done with filters, a common type being one octave wide. The noise environment, both outside and inside the studio, can be analyzed by determining the sound pressure levels within a series of 9 or 10 octave bands or within a greater number of half-octave or third-octave bands.* The word *level* implies a calibrated system in which the sound pressure levels are expressed in decibels above the standard reference level of 0.0002 microbar. Measurement of sound slices of the audible band this way yields the *spectrum* of the sound and tell us how the sound energy is distributed throughout the band of audible frequencies. By this system the spectrum of the interfering noise (upper curve in Fig. 12-3) may be determined as well as a background noise spectrum which can be tolerated in studio operation. A single-figure overall or *wideband* level of a street noise, for example, would be of limited value to us. We need to know the *spectrum* of the noise, and this requires a bit of specialized equipment.

MEASURING SOUND

Because equipment to measure sound levels is expensive to purchase or rent, and the interpretation of the results requires a certain amount of specialized training and experience, engaging a consultant may be the cheapest and most satisfactory course to follow. However, a general understanding of data required and methods used in obtaining

251

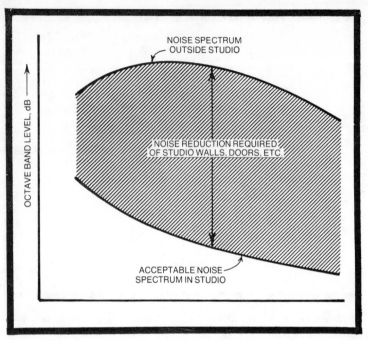

Fig. 12-3. The noise reduction which the studio barriers must supply is the difference between the outside noise level and the level acceptable within the studio. This varies with frequency.

it should be helpful from both the technical direction and the overall management point of view.

The simplest form of equipment capable of measuring the spectrum of a sound is a calibrated sound level meter and an associated set of filters. Figure 12-4 shows the General Radio Model 1933 sound level meter and analyzer. This precision instrument contains 10 bandpass filters one octave wide having center frequencies from 31.5 Hz to 16 kHz. The Brüel and Kjaer Type 2203 sound level meter and Type 1631 octave filter set shown in Fig. 12-5 is another such instrument. The B&K filter set has an additional octave centered on 31.5 kHz.

By selecting one filter after another, a series of readings are obtained which, when plotted, portray the spectral

---

* For most purposes, octave band measurements yield sufficient detail, but one can readily transpose data from one system to another. For instance, measured octave band levels can be transposed to equivalent half-octave band levels by subtracting 3 dB; to equivalent third-octave band levels by subtracting 5 dB. These corrections are approximate and are based on a continuous noise spectrum.

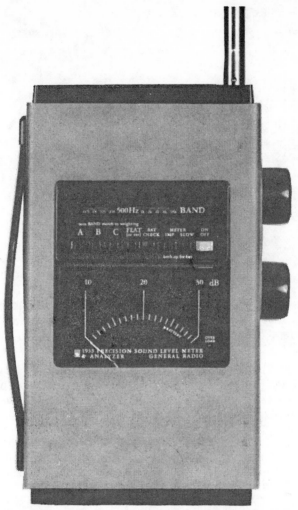

Fig. 12-4. Precision sound level meter equipped with an octave band filter set for analysis of noise, General Radio Model 1933.

signature of a given noise. Noise levels fluctuate, hence some sort of statistical approach must be used in gathering data.

Figure 12-6 displays the results of a serious study of freeway traffic noise.[1] The measuring instruments were set up 100 ft from the nearest edge of interstate highway I-94, five miles west of Detroit, a 70 mph speed limit prevailing at the time. The highest band levels encountered were in the octave bands centered on 63 and 125 Hz and generally lower in other

A

B

Fig. 12-5. In A, a precision sound level meter, Bruel & Kjaer Type 2203. B shows octave filter set for use with the B&K sound level meter.

bands. The maximum, median, and minimum graphs of Fig. 12-6 portray the range of noise levels generated by passing vehicles. The maximum levels are the ones to be considered seriously in regard to interference with studio operations. Some might wonder why anyone would locate a studio within 100 ft of a busy freeway. But freeways are snaking their way through metropolitan areas to such an extent that many studios will have to consider such noise, albeit at a somewhat greater distance, perhaps. We can adjust the spectra of Fig. 12-6 downward 5 or 6 dB for each doubling of the 100 ft distance. Thus, at 200 ft we would expect a maximum level in the 125 Hz band of about 92−93 dB at 400 ft about 86−88 dB.

ESTABLISHING NOISE CRITERIA

How high can background noise be in a studio before it threatens to become a problem? In one sense, the answer to this question depends upon the nature of the recording jobs to be handled in that particular studio. Any program material having pauses or low-level passages works closer to background noise levels and thus demands greater attention to background noise. But even rock music with its high levels has beginnings, endings, and occasional pauses or low-level

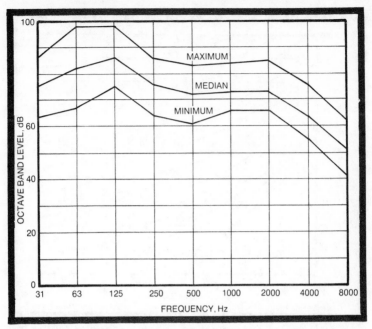

Fig. 12-6. Spectra of vehicular traffic noise 100 ft from a high-speed highway.[1]

intimate passages at times. A decision has to be made as to (1) how much quality—in the sense of quietness, and thus independence of outside conditions—one can afford; and (2) how much one can rely on electronic means such as noise gating to compensate for studio construction deficiencies. This is not an easy decision and it is further complicated by the excessive noise levels prevailing in many commercially successful studios operating today.

Being able to specify background noise spectra by single numbers would be a great convenience, and we can do this by standardizing the spectrum shape. The noise rating curves of Fig. 12-7, derived over a period of years on the basis of extensive experience, are now being considered for international adoption. Note that noise rating curve N-20 passes through 20 dB at 1 kHz, N-30 through 30 dB, etc. Note also that the curves have the general shape of the ear's sensitivity as determined by the corrected Fletcher—Munson characteristic. The higher level curves of Fig. 12-7 can serve as a guide with respect to annoyance and interference with speech under general circumstances. The lower ones are of more direct interest to us as they present possible practical goals for studio background noise which can be tolerated.

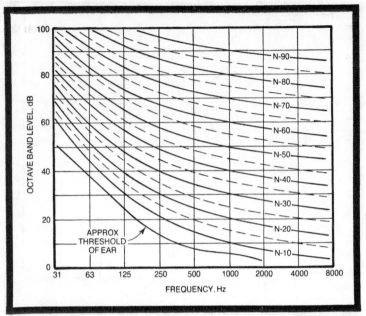

Fig. 12-7. Noise criteria rating curves proposed by the International Standards Organization. The levels given are for octave bands; for half-octaves the curves would be 3 dB lower; for third-octaves, about 5 dB lower.

The professionals generally specify noise levels between N-15 and N-25 as acceptable for broadcasting and recording studios. The N-15 curve is a suitable goal for radio drama studios and music recording studios in which quiet passages would allow background noise to be heard. If the music to be recorded is at consistently high levels, leaving only a ringing of the ears in what breaks occur, a higher studio background noise criterion can be selected. This would be a specialized studio, more or less unsuitable for recording speech and traditional music. In general, let us consider as a reasonable and practical goal a criterion of N-15 for a general-purpose recording studio.

It should be mentioned at this point that all of the above discussion and curves of Fig. 12-7 are for continuous, distributed spectrum noise. Any single-frequency components in the noise such as a whine or a whistle are extremely penetrating, as are transient and impulsive sounds such as doors slamming and footsteps on hard surfaces. A vehicular siren has a strong single-frequency component and, to make it even more difficult to guard against, may also be turned on and off or changed in pitch. Such instruments of torture have after all, been carefully designed to penetrate and to grab the attention of people.

BARRIER REQUIREMENTS

Once the noise levels are determined for the area being considered for the studio, and after a background noise criterion within the studio has been decided upon, the basic information is at hand for arriving at requirements for walls, floors, ceilings, and doors (assuming you are not considering windows). Our major source of noise, let us say, is traffic on a freeway 400 ft from the studio location. The maximum traffic noise spectrum of Fig. 12-6 is reduced 10 dB point by point (2 distance doubles at 5 dB each) and replotted in Fig. 12-8. Noise rating curve N-15 is also replotted. If the interfering sound outside is to be reduced to the N-15 level within the studio, the shaded area between the two graphs is the transmission loss required for the various barriers.

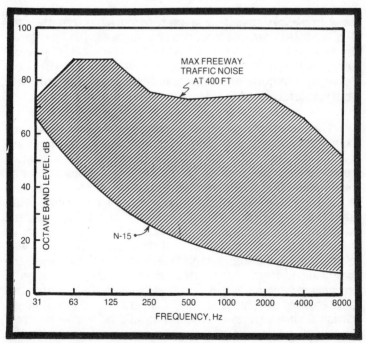

Fig. 12-8. The upper graph is the octave band levels for maximum freeway traffic noise adjusted for 400 ft distance. The lower graph is the N-15 contour accepted for the studio noise level allowable. The shaded area is the transmission loss which barriers must provide.

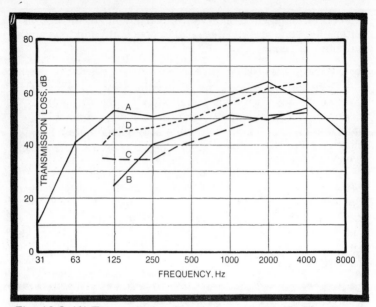

Fig. 12-9. (A) Transmission loss (TL) required; (B) TL offered by staggered-stud frame construction (STC 42); (C) TL offered by 8 in hollow concrete block wall (STC 45); (D) TL offered by 8 in concrete block wall filled solid with concrete and plastered on both sides (STC 56).

The transmission loss requirements (the difference between the two graphs at each octave center frequency) are plotted as in Fig. 12-9. We notice that the isolation requirements are greatest in the midband frequencies and tend to fall off at both extremes. In this middle region we find we need 50 to 65 dB transmission loss in our barriers. What kind of a wall will provide such a loss? Without distinguishing at the moment between exterior walls and inner walls, let us consider what several typical wall structures offer in the way of transmission losses. Graph B of Fig. 12-9 is for one of the best walls of frame construction available: 2-by-4 studs on 16 in. centers set on 2-by-6 plate, staggered so that the two plasterboard diaphragms are independent of each other. Graph C shows transmission loss of an 8 in. hollow concrete block wall, while graph C shows the increased loss obtained by filling the B wall with concrete and plastering both faces. None of these walls yields a transmission loss large enough to meet our requirements; so some form of multiple construction is obviously needed, whether it be a wall of two or more tiers or a room within a room.

In general, the heavier the walls, the greater the transmission loss. The so-called *mass law* is an expression of this. However, factors other than mass can come into play, causing a structure to have more transmission loss than its mass alone would seem to justify. The staggered-stud wall is a good example of this. Wall D in Fig. 12-9 weighs 67 lb for each square foot of wall surface and wall C 33.5 lb sq ft. Wall B, the staggered-stud wall, weighs only about 7 lb/sq ft and yet its transmission loss equals or exceeds that of wall B for most of the band. This results from isolating the wall diaphragm of one side from that of the other side of the wall and providing absorption in the space between.

While considering transmission loss as a function of frequency (as we have done in Fig. 12-9) is helpful for an understanding of what is going on, it is too unmanageable to carry any further. What we need is a single number rating which tells us what we need to know to define transmission loss characteristics of a given wall. Averaging decibels of transmission loss for the different frequencies raises some real theoretical questions, although this was actually the first approach to a single-number system. This has been superseded by the *sound transmission class* (STC) concept which is intended to achieve a more convenient and realistic averaging. In Fig. 12-9 the STC ratings of the three walls considered are shown; such ratings will be used extensively in the following discussion.

Concrete Walls for Noise Control

If mass is what we need for high-transmission-loss barriers (and it is, with certain reservations), the first material we think of is concrete. It is both dense and inexpensive. For new construction, concrete walls can meet structural, esthetic, and acoustical specifications.

In Fig. 12-10 we see that a 4 in. wall of concrete gives a transmission loss of about STC 48 dB. Doubling the thickness increases this only to 52 dB. This is a substantial transmission loss but insufficient, in itself, to shield our studio from the freeway traffic noise at 400 ft, as we saw in Fig. 12-8. However, concrete is ideal for the outer shell of buildings and for partitions, roofs, and floors *if there is adequate provision for carrying the load.*

Concrete Masonry Walls for Noise Control

Concrete block construction is popular because of its low cost. Concrete block partitions are quite effective in shutting out unwanted noise. Figure 12-10 compares three types of

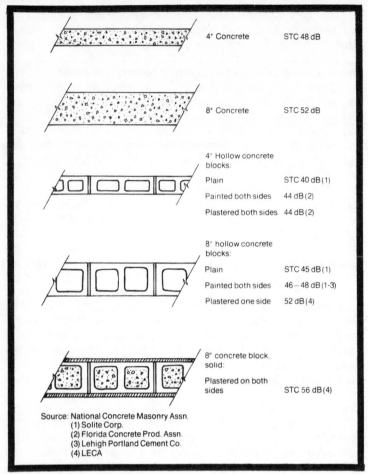

	4″ Concrete	STC 48 dB
	8″ Concrete	STC 52 dB
	4″ Hollow concrete blocks:	
	Plain	STC 40 dB (1)
	Painted both sides	44 dB (2)
	Plastered both sides	44 dB (2)
	8″ hollow concrete blocks:	
	Plain	STC 45 dB (1)
	Painted both sides	46 – 48 dB (1-3)
	Plastered one side	52 dB (4)
	8″ concrete block, solid:	
	Plastered on both sides	STC 56 dB (4)

Source: National Concrete Masonry Assn.
    (1) Solite Corp.
    (2) Florida Concrete Prod. Assn.
    (3) Lehigh Portland Cement Co.
    (4) LECA

Fig. 12-10. Transmission loss of representative types of concrete wall construction.

concrete block structures. The $4 \times 8 \times 16$ in. block gives a wall 4 in. thick and a rating of STC 40 dB. Painting or plastering increases this only a few decibels. By using $8 \times 8 \times 16$ in. blocks, a wall of 8 in. thickness and an STC rating of 45 dB results. Painting both surfaces gives only a very marginal gain, but plastering on only one side gives a 7 dB increase in transmission loss over the plain surfaces. By plastering both surfaces of this 8 in. block wall and filling the voids with well-rodded concrete, the STC rating goes up to 56 dB. An intermediate STC rating can be obtained by filling the voids with sand. There are many more configurations and resulting

STC values of block and concrete walls which can be found in publications of the National Concrete Masonry Association and other sources.[2,3,4]

Frame Walls For Noise Control

Walls framed of lumber, gypsum board, etc. are often the best for inside partitions, both for new construction and renovation jobs. For one thing, a frame wall yielding a given transmission loss weighs much less than either a poured concrete or concrete block wall. This follows from the fact that principles other than the brute force mass law are involved in such walls. Referring to Fig. 12-11 we can see that a staggered-stud wall with insulation gives an STC of 49 dB with a density of 7.2 lb/sq ft, which is quite close to the STC rating of 52 dB for the 8 in. concrete wall of Fig. 12-10 weighing 96 lb/sq ft. A double-stud wall filled with insulation gives one of the best walls available (STC 58 dB) short of going to very massive or more complex multileaved structures.

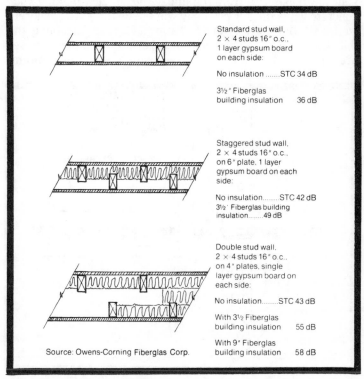

Standard stud wall,
2 × 4 studs 16″ o.c.,
1 layer gypsum board
on each side:

No insulation ....STC 34 dB

3½″ Fiberglas
building insulation   36 dB

Staggered stud wall,
2 × 4 studs 16″ o.c.,
on 6″ plate, 1 layer
gypsum board on each
side:

No insulation.......STC 42 dB
3½′ Fiberglas building
insulation.......49 dB

Double stud wall,
2 × 4 studs 16″ o.c.,
on 4″ plates, single
layer gypsum board on
each side:

No insulation.......STC 43 dB

With 3½ Fiberglas
building insulation   55 dB

With 9″ Fiberglas
building insulation   58 dB

Source: Owens-Corning Fiberglas Corp.

Fig. 12-11. Transmission loss of representative types of frame wall construction.

The concrete, concrete masonry, and frame barriers considered to this point give adequate transmission loss for many applications, but what they can do in acoustic isolation is limited. Situations are often encountered in which the airborne and impact noises are so great as to create serious problems for a recording studio. For example, let us say that a certain location is very desirable from every point of view but one: the clicking of heels on the hard concrete floor of a nearby pedestrian mall. Such impact sounds are very difficult to control. They are transmitted great distances through the solid structure of the building with practically no loss. Not only that, but they can set a concrete or masonry wall, floor, or ceiling some distance away vibrating as a diaphragm, radiating appreciable sound energy into the air in a studio.

A good rule to follow is to try first to reduce an offending noise at the source. If it is an indoor corridor or mall, a resilient floor covering may be the cheapest answer. If this seems to be impractical, you must take heroic steps to protect your low-noise area from the impact noise. Just making your concrete walls thicker probably will not do the job. A concrete wall 8 in. thick gives an STC rating of 52 dB; doubling the thickness to 16 in. gains only about 5 dB. If the impact noises are being radiated from a wall active as a diaphragm,

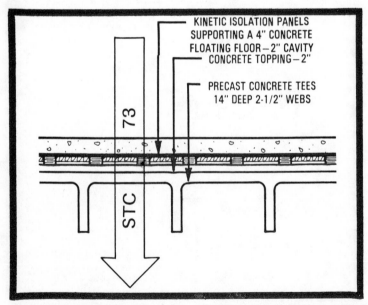

Fig. 12-12. Transmission loss of typical floating floor construction. (Consolidated Kinetics Corp.)

increasing the thickness of the wall may help some, but is basically the wrong approach to a solution.

THE FLOATING PRINCIPAL

Making the walls, floor, and ceiling a composite structure offers real promise in cases in which extra protection from outside noise is needed or for protecting persons outside from sounds generated within. It is just as accurate to refer to composite walls and composite ceilings as floating walls and floating ceilings. The floating or composite structures to be described can salvage difficult situations and make possible the location of studios in noisy but otherwise favorable locations. Resiliently supported partitions, ceilings, and floors offer sound transmission losses up to 10 dB greater than can be achieved by mass alone.

Floating the Floor

An exceptionally good floating floor construction yielding STC 73 dB is shown in Fig. 12-12. This floor is a 4 in. concrete slab poured on top of isolation panels such as shown in Fig. 12-13. These panels carry both the live and dead load of the

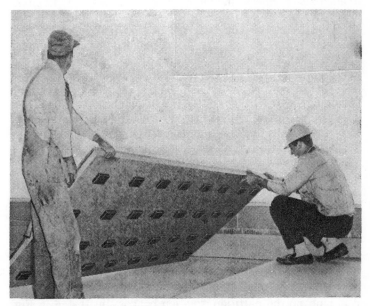

Fig. 12-13. The effectiveness of the floating floor is determined by the vibration pads used. The isolation pads pictured are made of special glass fiber. (Consolidation Kinetics Corp.)

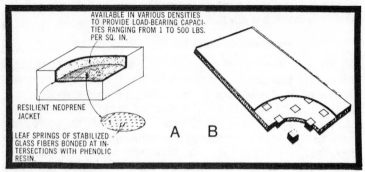

Fig. 12-14. Details of fabrication of vibration isolation pads and the plywood or hardboard isolation panels. (Consolidated Kinetics Corp.)

floor to the structural floor beneath through *vibration isolation* pads. These pads may be of rubber or other resilient material carefully designed to carry the load. Consolidated Kinetics Corporation[5], whose techniques will be closely followed in the ensuing discussion, offers pads of precompressed, molded-glass fibers individually coated with a flexible, moisture-impervious membrane. Various pad densities and spring rates are used for bearing loads of 1 to 500 lb/sq in., as required. Figure 12-14A shows the construction of individual pads and Fig. 12-14B shows how they are factory bonded to the floor isolation panels of plywood or hardboard. Low-density fiberglass is bonded to the entire area of the panel between the isolation pads to further reduce the transmission of sound. An interesting feature is that the natural period of vibration of these panels (about 12 Hz), and hence their isolating ability, is independent of load. These isolation panels can sustain a 300% overload without damage.

The floating floor is isolated from the structural floor as described above. It must also be decoupled from walls, columns, etc. This is accomplished by the use of perimeter isolation board such as shown in Fig. 12-15A. This board is also of molded fiberglass coated with a flexible moisture-impervious membrane.

Floating Partitions

To decouple a partition from the floor upon which it rests, it is built on the partition-isolation boards illustrated in Fig. 12-15C. These are ⅛ in. hardboard bonded to and supported by 7/16 in. thick (10 lb/cu ft) molded fiberglass coated with a flexible moisture proof membrane. These panels are designed to carry partition loads up to 1500 lb/ sq ft and have reserve

264

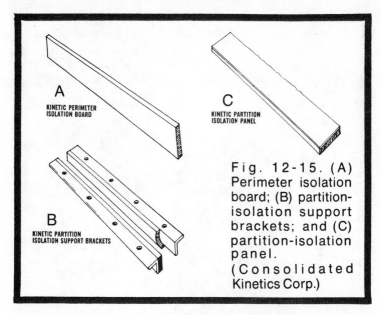

A
KINETIC PERIMETER
ISOLATION BOARD

B
KINETIC PARTITION
ISOLATION SUPPORT BRACKETS

C
KINETIC PARTITION
ISOLATION PANEL

Fig. 12-15. (A) Perimeter isolation board; (B) partition-isolation support brackets; and (C) partition-isolation panel. (Consolidated Kinetics Corp.)

capacity of three times this amount. If the top of the partition were solidly mounted to the slab above, the benefit of having the partition riding on resilient board below would be canceled. To float the partition fully, it stops just short of the upper slab surface and is kept in an upright position by the partition-isolation support brackets of Fig. 12-15B. These brackets are angles of structural steel to which is bonded fiberglass covered with a membrane. These brackets not only restrain the top of the partition, but provide an airtight acoustic seal. The partition is also decoupled from adjacent walls and columns by the insertion of perimeter isolation board. If a double partition is used, the cavity between is filled with low-density fiberglass.

Floating the Ceiling

A floating floor and floating walls and partitions leave us vulnerable from above. As Paul Simon so lyrically phrased it, "one man's ceiling is another man's floor." The structural slab constituting the ceiling may be a floor on which hard heels beat their tattoo and which carry vibrations from traffic, elevators, air-conditioning machinery and the like. If the ceiling is next to the roof, noise from aircraft can also be a serious source of interference. The answer to such problems may be in a suspended ceiling. The usual integrated or suspended ceiling of lightweight fiberboard supported by wires

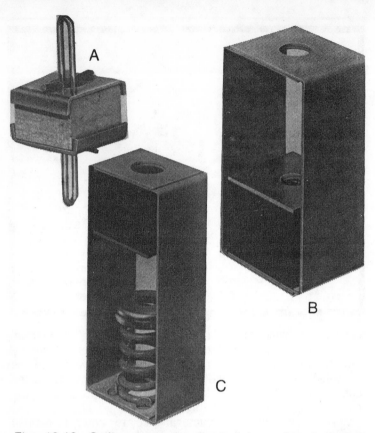

Fig. 12-16. Ceilings can be floated by using isolation hangers: (A) glass fiber element (B) for loads up to 2700 lb, and (C) hanger combining glass fiber and spring. (Consolidated Kinetics Corp.)

does not offer much help. A specially designed suspended ceiling, isolated properly from the supporting slab, can provide up to 10 dB more sound transmission loss than the direct-tied ceiling. The secret is in having a heavier ceiling and supporting it with isolation hangers of the type shown in Fig. 12-16. Ceilings having a mass of 5 to 15 lb/sq ft and appropriate hangers are called for when this isolation is required. These ceilings may be of gypsum board or lath-and-plaster, and should be sealed properly around the periphery.

It is emphasized that the isolation hangers must be carefully matched to the ceiling load to be carried and to the characteristics of the sound from which protection is desired. Figure 12-16A shows a simple device designed to isolate ceilings from vibrations of the building structure. It depends on the resiliency of a fiberglass element for its effect and can be used for loads up to 200 lb. Figure 12-16B is a heavier unit designed for greater permanency and commonly used to support mechanical equipment as well as heavy ceilings or other loads up to 2700 lb. The hanger of Fig. 12-16C combines, in series, a fiberglass pad which provides good isolation from the higher audible frequency vibrations and a spring which is more effective against the lower frequencies. Such a unit can be used to advantage in protecting a quiet area from sounds of overflying aircraft. They are rated to 2600 lb. All of these can carry 500% overload without failure and are of failsafe construction.

Applying the Floating Principle to Studios

In Fig. 12-17 the floor plan of a suite of three studios and supporting areas is shown. It is assumed that the large studio

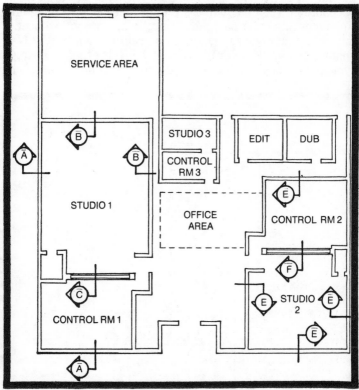

Fig. 12-17. Floor plan of typical suite of studios and supporting areas. Sections indicated are shown in following figures.

1, the medium studio 2, and the small studio 3 are to provide a diversity of services and, for this reason, have different requirements as to background noise levels which can be tolerated within. Let us say that studio 1 requires the maximum of isolation, studio 2 next, and studio 3 least. Studio 1 and 2 have walls shared with neighboring commercial and professional activities. What kind of floors, walls, and ceilings should be specified for each? To answer this accurately for any specific situation requires much more thought and planning than this example suggests, but to illustrate when and how the floating principle should be applied in studio design we shall jump headlong into this hypothetical and necessarily arbitrary example.

The west wall of studio 1 is a shared wall and we must protect this neighbor from the high sound levels generated in the studio and control room as well as protect ourselves from outside noises he might make. This illustrates the wide range of conditions required to make a studio general in its functions. For recording quiet passages of music or speech, we need a low background noise level such as represented by the N-15 contour of Fig. 12-7. For the very high levels created during a typical rock session, background noise exceeding N-15 would

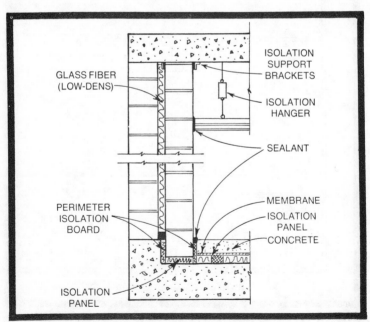

Fig. 12-18. Standing inner wall isolated at top and bottom and isolated ceiling and floor.

be no problem, but the walls of high transmission loss may still be needed to avoid being a nuisance to the neighbors. Monitoring loudspeaker levels of 120 dB (we'll consider this an absolute maximum, since it overlaps the threshold of physical pain!) in the control room would be reduced to only 75 dB on the neighbor's side of the common wall with an 8 in. concrete block wall offering 45 dB transmission loss. A level of 75 dB is a comfortable level if you *want* to listen to music; if you don't, it can be a class A irritant. To meet this dual need before complaints come in, let us specify a west wall as shown in Fig. 12-18; this is designated **A** in Fig. 12-17. The outside concrete block wall is augmented by a similar inner concrete block wall isolated completely around its periphery by the special isolation boards previously described. In this sectional view we also see how the suspended ceiling and floating floor are related to this composite wall structure.

The north and east walls of studio 1 (designated **B**) are interior partitions. Here again we need to protect studio 1 from normal activity noises within the building (a 75 dB level escaping from studio 1 would certainly be intolerable to other studio functions). For this reason, **B** will be a continuation of the double-wall construction, as shown in Fig. 12-19.

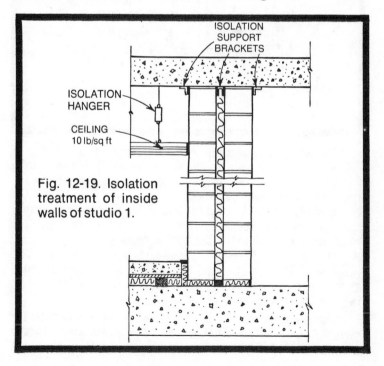

Fig. 12-19. Isolation treatment of inside walls of studio 1.

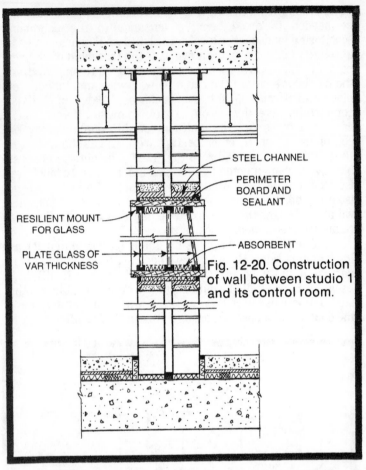

Fig. 12-20. Construction of wall between studio 1 and its control room.

STEEL CHANNEL

PERIMETER BOARD AND SEALANT

RESILIENT MOUNT FOR GLASS

PLATE GLASS OF VAR THICKNESS

ABSORBENT

The wall between studio 1 and its control room is also of double construction (C), as shown in Fig. 12-20. We must take some extra precautions if the acoustic integrity of this wall is not to be compromised by the observation window. Perimeter board of Fig. 12-15A is used to isolate the window frame from the channel iron opening. This is thoroughly sealed after packing. A triple glass window will be required.

The sectional views have shown wall isolation arrangements at the top and the bottom of the walls. Figure 12-21 shows plan views of how the same walls are isolated on the vertical edges.

The requirements for studio 2 are less stringent than for studio 1. Let us say that a careful survey leads to the conclusion that floating floor, walls, and ceiling are not

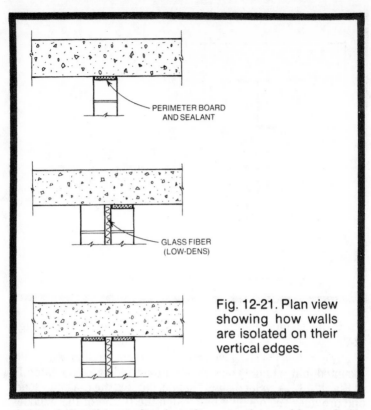

PERIMETER BOARD
AND SEALANT

GLASS FIBER
(LOW-DENS)

Fig. 12-21. Plan view showing how walls are isolated on their vertical edges.

required for this studio, but there is the problem of the high-level sounds generated in it and its control room bothering neighbors as well as other studio activities. We are relaxing the N-15 requirements for studio 2 on the basis that one studio meeting such a low background noise criterion is enough for one organization, and is all that this organization can afford. Measurements made in the building at different times of day give hope that the N-20 criterion may be at least approached by increasing the transmission loss of the walls of studio 2. As the primary use of this studio is mixdown and recording of rock bands, the high internal sound levels also indicate heavier walls. A study of the structure-borne exposure to noise tells us that we can stop worrying about that. We need not be concerned about tenants below because the studio is on the ground level and there are none. There is some chance of troublesome noise from the offices above, but not enough to justify floating ceilings and walls. Considering all the pertinent data, we decide on increasing the transmission loss of the walls by employing a double concrete block wall.

271

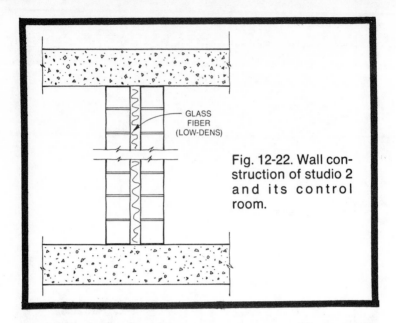

GLASS
FIBER
(LOW-DENS)

Fig. 12-22. Wall construction of studio 2 and its control room.

The walls designated **E** of Fig. 12-17 are thus described by Fig. 12-22.

The wall separating control room 2 from studio 2 will have mounted in it a triple glass observation window (designated **F** in Fig. 12-23). A welded-steel channel carries the masonry load and provides the opening into which the frame of the window fits. The space between the channel and the wood or other frame is filled with perimeter isolation board and sealed at the outer edges.

We shall do nothing unusual as far as studio 3 is concerned. It is a speech studio and ordinary masonry or frame walls giving an STC of around 45 or 50 dB should serve well in isolating the studio and control room from outside noises and isolating the studio from the control room.

All of the numerous factors we have been considering for isolating floors, walls, and ceilings are pulled together in Fig. 12-24 for single masonry walls and in Fig. 12-25 for the lighter metal stud and drywall construction.

The Floating Principle in the Machine Room

Figures 12-26 and 12-27 illustrate in considerable detail some time-proved methods of keeping machinery noises confined largely to the machine room. It takes good engineering and meticulous attention to detail during construction and installation to control such noises and

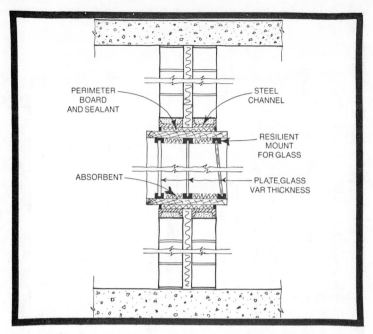

Fig. 12-23. Construction of wall separating studio 2 and its control room.

In the figure the following labels appear: PERIMETER BOARD AND SEALANT, STEEL CHANNEL, RESILIENT MOUNT FOR GLASS, ABSORBENT, PLATE GLASS VAR THICKNESS.

vibrations. All isolation systems must be free of rigid connections to the building structure. All openings must be properly engineered; an acoustical engineer should evaluate the minimum transmission loss required and other critical requirements.

ACOUSTIC DOORS

No sound barrier is better than its weakest part and the effect of good, solid walls can easily be nullified by inserting poor doors or windows in them. The mass law applies to doors as well as walls—the heavier the door, the better its sound transmission loss. Common household doors are made of thin panels enclosing hollow spaces and are invariably poorly sealed; consequently, they offer little protection against outside noise. Doors of solid wood are much better and the do-it-yourselfer can glue up quite a respectable door by laminating three sheets of ¾ in. particle board, which is superior to plywood because of its greater density.

Composite walls, we have seen, are able to beat the mass law; that is, they offer greater transmission loss than their surface density would entitle them to on a straight-mass basis.

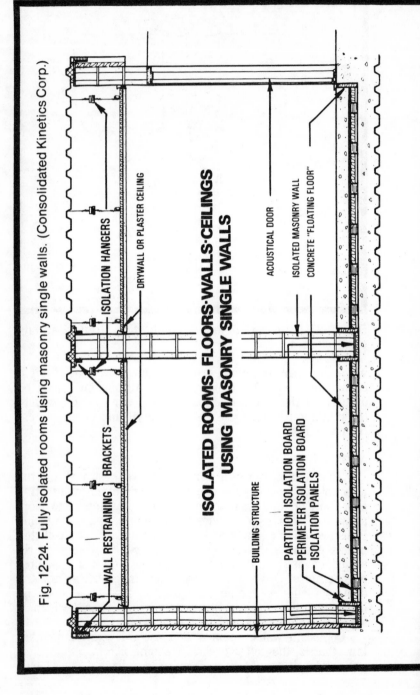

Fig. 12-24. Fully isolated rooms using masonry single walls. (Consolidated Kinetics Corp.)

ISOLATED ROOMS- FLOORS-WALLS-CEILINGS
USING MASONRY SINGLE WALLS

ISOLATION HANGERS

DRYWALL OR PLASTER CEILING

ACOUSTICAL DOOR

ISOLATED MASONRY WALL

CONCRETE "FLOATING FLOOR"

WALL RESTRAINING BRACKETS

BUILDING STRUCTURE

PARTITION ISOLATION BOARD
PERIMETER ISOLATION BOARD
ISOLATION PANELS

274

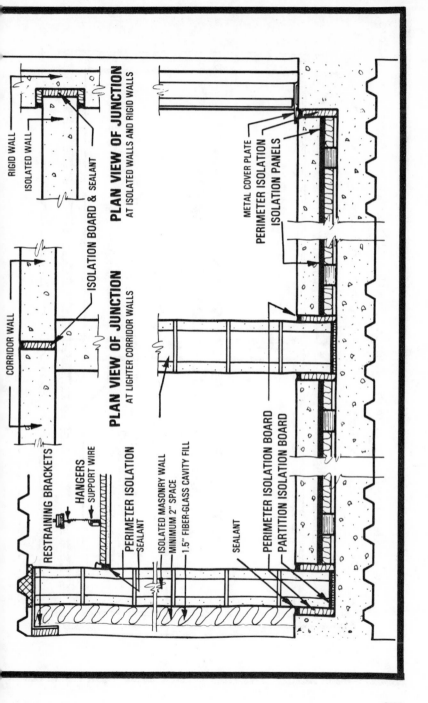

PLAN VIEW OF JUNCTION
AT ISOLATED WALLS AND RIGID WALLS

RIGID WALL

ISOLATED WALL

ISOLATION BOARD & SEALANT

PLAN VIEW OF JUNCTION
AT LIGHTER CORRIDOR WALLS

CORRIDOR WALL

ISOLATION BOARD

METAL COVER PLATE

PERIMETER ISOLATION

ISOLATION PANELS

RESTRAINING BRACKETS

HANGERS
SUPPORT WIRE

PERIMETER ISOLATION
SEALANT

ISOLATED MASONRY WALL
MINIMUM 2" SPACE

1.5" FIBER-GLASS CAVITY FILL

SEALANT

PERIMETER ISOLATION BOARD
PARTITION ISOLATION BOARD

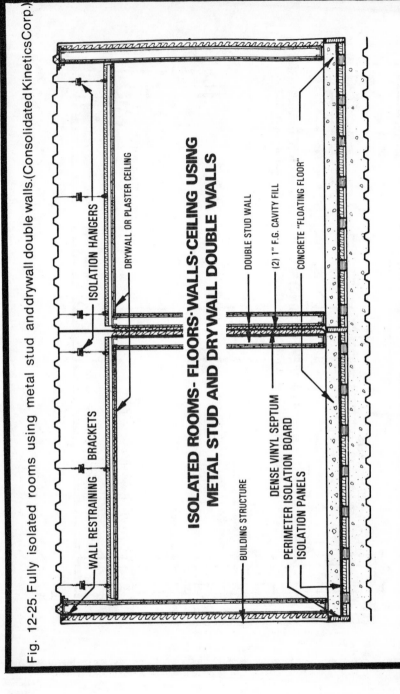

Fig. 12-25. Fully isolated rooms using metal stud and drywall double walls. (Consolidated Kinetics Corp.)

ISOLATED ROOMS—FLOORS·WALLS·CEILING USING
METAL STUD AND DRYWALL DOUBLE WALLS

ISOLATION HANGERS

DRYWALL OR PLASTER CEILING

DOUBLE STUD WALL

(2) 1" F.G. CAVITY FILL

CONCRETE "FLOATING FLOOR"

WALL RESTRAINING

BRACKETS

BUILDING STRUCTURE

DENSE VINYL SEPTUM

PERIMETER ISOLATION BOARD

ISOLATION PANELS

276

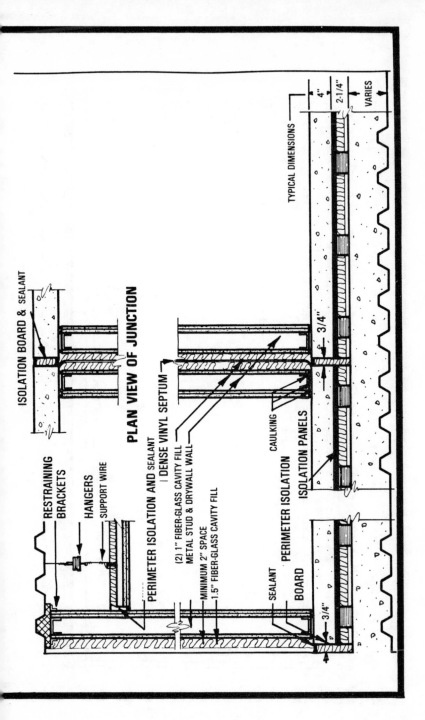

ISOLATION BOARD & SEALANT

PLAN VIEW OF JUNCTION

RESTRAINING BRACKETS

HANGERS

SUPPORT WIRE

PERIMETER ISOLATION AND SEALANT

DENSE VINYL SEPTUM

(2) 1" FIBER-GLASS CAVITY FILL

METAL STUD & DRYWALL WALL

MINIMUM 2" SPACE

1.5" FIBER-GLASS CAVITY FILL

CAULKING

SEALANT

PERIMETER ISOLATION BOARD

ISOLATION PANELS

TYPICAL DIMENSIONS

4"

2-1/4"

VARIES

3/4"

3/4"

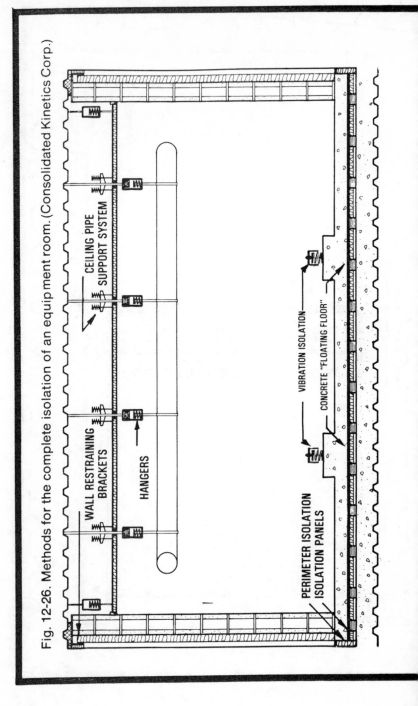

Fig. 12-26. Methods for the complete isolation of an equipment room. (Consolidated Kinetics Corp.)

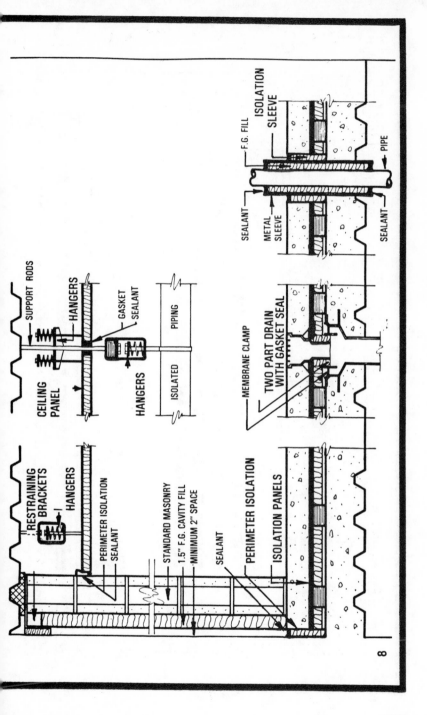

SUPPORT RODS

HANGERS

GASKET
SEALANT

CEILING PANEL

PIPING

ISOLATED

HANGERS

RESTRAINING BRACKETS

HANGERS

PERIMETER ISOLATION

SEALANT

STANDARD MASONRY

1.5" F.G. CAVITY FILL

MINIMUM 2" SPACE

SEALANT

PERIMETER ISOLATION

ISOLATION PANELS

MEMBRANE CLAMP

TWO PART DRAIN WITH GASKET SEAL

ISOLATION SLEEVE

F.G. FILL

SEALANT

METAL SLEEVE

PIPE

SEALANT

8

279

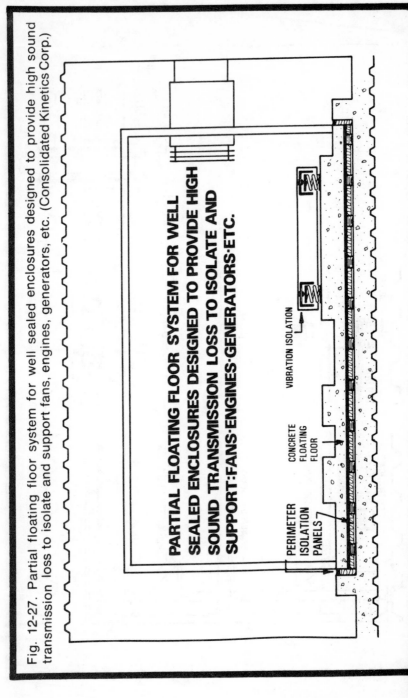

Fig. 12-27. Partial floating floor system for well sealed enclosures designed to provide high sound transmission loss to isolate and support fans, engines, generators, etc. (Consolidated Kinetics Corp.)

PARTIAL FLOATING FLOOR SYSTEM FOR WELL SEALED ENCLOSURES DESIGNED TO PROVIDE HIGH SOUND TRANSMISSION LOSS TO ISOLATE AND SUPPORT:FANS·ENGINES·GENERATORS·ETC.

PERIMETER ISOLATION PANELS

CONCRETE FLOATING FLOOR

VIBRATION ISOLATION

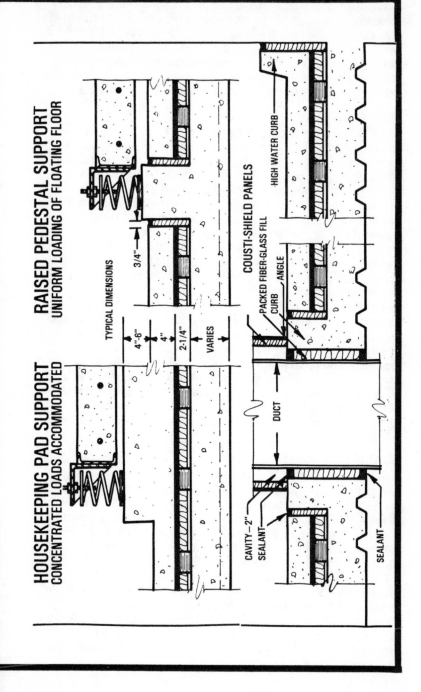

RAISED PEDESTAL SUPPORT
UNIFORM LOADING OF FLOATING FLOOR

HOUSEKEEPING PAD SUPPORT
CONCENTRATED LOADS ACCOMMODATED

TYPICAL DIMENSIONS

3/4"

4"-6"

4"

2-1/4"

VARIES

COUSTI-SHIELD PANELS

HIGH WATER CURB

PACKED FIBER-GLASS FILL

CURB ANGLE

DUCT

CAVITY—2"

SEALANT

SEALANT

Composite doors can do the same. Commercially available acoustic doors[6] quite generally rely on this principle. Transmission losses in the 40–45 dB range are achieved in such doors if great care is exercised in treatment of the peripheral seal. There are many products on the market today such as lead and lead-loaded vinyl, which can serve well in reducing the transmission of sound through a composite door.

SOUND LOCK

If you have a 50 or 55 dB wall, it will be difficult matching it with a single door; and even if you do, the resulting door will be heavy to swing and difficult to seal. Further, some door seals have the irritating faculty of deteriorating with use and passage of time while the transmission loss of the wall remains stable. A far more practical method of making doors match the walls acoustically is to have two modest doors in series in a *sound lock* or vestibule arrangement. Two 36 dB doors separated at least three feet with a highly absorbent space between can achieve a 50 dB overall transmission loss inexpensively and conveniently.[7] Opening a single door to a studio momentarily subjects the studio to the full force of the noise prevailing outside. With a sound lock, a person can first enter the vestibule between the two doors and then open the inner door with negligible disturbance to those working in the studio. A three-door sound lock can serve as a noise lock between studio, control room, and the outside corridor. The walls of the sound lock space should be treated with efficient wide-range absorbers and the floor covered with carpet to reduce the vestibule noise levels. It is also advisable to treat the outer corridor to reduce the noise level at the outside door of the sound lock.

SEALING ACOUSTIC DOORS

Doors with high transmission loss can be very heavy to swing and latch, yet their effectiveness depends upon a tight seal with the door frame. There are many methods for sealing doors—some good and some ineffective and temperamental. One of the most satisfactory is the magnetic door seal. These seals are commonly used on refrigerator doors, particularly the larger meat-storage type. They have been used extensively by the BBC with great success. They have the advantages sought in all studio door seals of being able to seal effectively even if the door warps somewhat with time, and to be relatively easy to open and tending to be self-closing as the door approaches its seated position. With the magnetic seals, pneumatic door closers operate satisfactorily and fire regulations are more easily met.

A very promising proprietary seal[8] is pictured in Fig. 12-28. This noise-barrier gasket features a three-fingered design that functions as three airlocks to increase sound transmission loss through door openings. It is made of tough extruded polypropylene that withstands the continued flexing and wiping action of normal use. It is claimed that this gasket material, properly installed, reduces sound transmission an amount approaching that of the same door permanently sealed along the edges with five beads of nonhardening calking! Noise reduction of 51 dB has been measured under laboratory conditions on a 1.75 in. door weighing 6.2 lb/sq ft.

OBSERVATION WINDOWS

It is well to summarize the basic characteristics we must work for to achieve observation windows of high transmission loss:

- In a framed staggered-stud or double wall, mount one glass plate to one wall and the other plate to the other wall; do not connect the two walls rigidly in any way.
- Use glass plates of heavy weight and different thicknesses.
- All glass plates should be resiliently supported around their periphery.
- The reveals between the glass plates should be made absorbent.
- Glass fiber packing and calking compound should be generously used to insure airtight joints between frame and wall.
- Inclining one glass plate from the vertical may be done to minimize light reflections and sound reflection problems in the studio. This has negligible effect on transmission loss.
- Minimum spacing of glass plates of 4 or 5 in. is recommended.

Preassembled double-glass windows are available commercially. The use of extruded aluminum and molded Neoprene insures sealing and performance considerably above the mass law.

AIR-CONDITIONING NOISE

In heating, ventilating, and air-conditioning systems,[9] the primary noise source is the fan that moves the air through the ducts. Noise can be introduced to the quiet area either through these ducts or via structure-borne paths. Noise may also be generated by the turbulence resulting from high-velocity air moving around unstreamlined objects. A damper in a duct is a

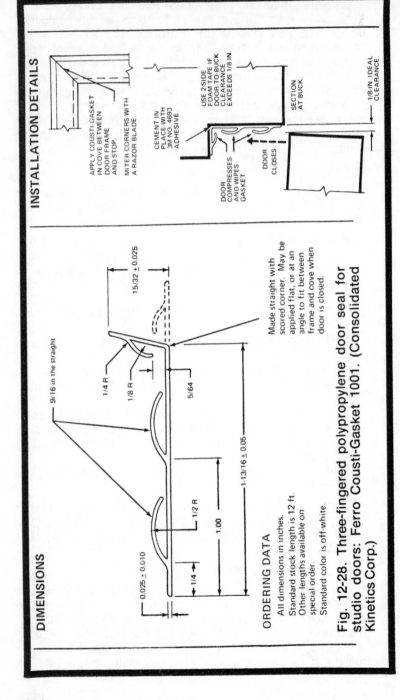

INSTALLATION DETAILS

APPLY COUSTI-GASKET IN COVE BETWEEN DOOR FRAME AND STOP.

MITER CORNERS WITH A RAZOR BLADE.

CEMENT IN PLACE WITH 3M NO. 4693 ADHESIVE.

USE 2 SIDE FOAM TAPE IF DOOR-TO-BUCK CLEARANCE EXCEEDS 1/8 IN.

SECTION AT BUCK.

1/8-IN. IDEAL CLEARANCE.

DOOR COMPRESSES AND WIPES GASKET

DOOR CLOSES.

DIMENSIONS

15/32 ± 0.025

9/16 in. the straight

1/4 R

1/8 R

5/64

1-13/16 ± 0.05

1.00

1/2 R

1/4

0.025 ± 0.010

Made straight with scored corner. May be applied flat, or at an angle to fit between frame and cove when door is closed.

ORDERING DATA

All dimensions in inches.
Standard stock length is 12 ft. Other lengths available on special order.
Standard color is off-white.

Fig. 12-28. Three-fingered polypropylene door seal for studio doors: Ferro Cousti-Gasket 1001. (Consolidated Kinetics Corp.)

good example of a generator of noise through turbulence. Another example is a poorly designed grille and diffuser.

If major silencing of fan noise is required, there are several basic forms such silencers can take.[10] In Fig. 12-29A a metal duct is simply lined with sound absorbing material, commonly glass fiber board. There are ducts available which are made of glass fiber material, delivered flat and folded into rectangular duct form and taped on the job. Sound traveling down a duct will have some of its energy in transverse modes, some in plane-wave modes. The lining acts more effectively on the transverse mode. This means that the transverse mode may die out within a few feet and the plane-wave mode traveling down the duct is relatively less affected by the lining.

The lined bend of Fig. 12-29B is more effective on this plane-wave mode, as the sound must undergo multiple reflections to negotiate the bend. The reactive chamber of Fig. 12-29C reflects some energy back toward the source much as a discontinuity in a radio-frequency transmission line. The plenum chamber of Fig. 12-29D combines the effects of A, B, and C. The tuned stub of Fig. 12-29E, absorbing in a narrow frequency band, may be a useful tool in combating a particularly obnoxious single-frequency component of the noise traveling down the duct.

Some designers prefer to enclose the fan in a plenum chamber along with air filter, heat transfer elements, and humidifying and dehumidifying equipment. Some of the thermal insulation required can be applied as acoustic treatment of the inner surfaces, making the fan plenum serve also as an acoustic plenum. It is possible to reduce noise levels 10—20 dB in the usually critical 125 Hz octave band by employing this system. As fan noise is typically the loudest and of greatest concern in a low-pressure system, silencing close to the fan discharge may provide all the silencing necessary in the usual installation.

Silencing further downstream is normally required to meet the extreme demands in radio, television, and recording studios. One reason for this is that the ducts are a transmission path for sound to travel from one studio to another. Proper arrangement of ducts can minimize this problem. Lining the ducts with absorbent material is a further step in the right direction. In extreme cases, sound baffles may be required. Figure 12-30 illustrates some precautions that should be observed. In Fig. 12-30A sound readily passes from one studio to its control room or to another studio because of the short path afforded by the duct arrangement. Lining the duct will somewhat reduce the coupling between rooms, but probably

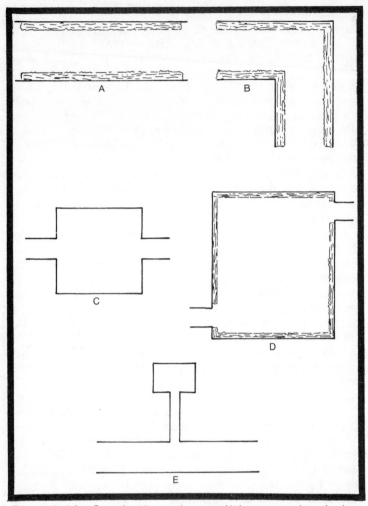

Fig. 12-29. Combating air-conditioner noise being transmitted through the duct by: (A) lining a straight duct with absorbing material; (B) lining a bend in a duct; (C) reactive chamber; (D) lined plenum chamber; and (E) tuned stub.

not enough. Locating the ducts as in Fig. 12-30B is far better in regard to isolating one studio from the other; again, lining the common duct would increase this isolation modestly. If both studios must be served by a common duct and the arrangement of Fig. 12-30B is marginal, placing a few lined right-angle bends in the duct between the two studios, as in

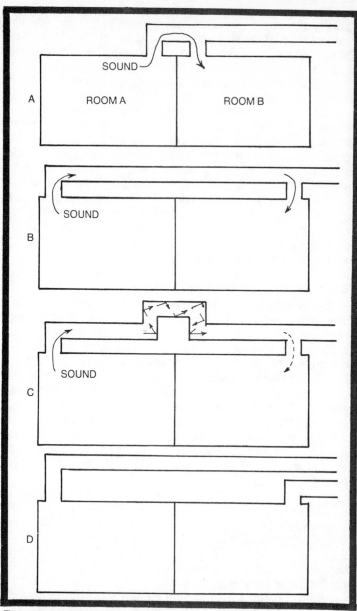

Fig. 12-30. Combating sound traveling from one room to another through ducts: (A) worst case (resulting from short path); (B) improvement due to lengthened duct; (C) further improvement, lined bend; and (D) greatest isolation by feeding each room with its own duct.

sketch C, would probably solve the problem. Ideally, maximum isolation between studios, as far as sound traveling through the duct is concerned, is obtained by serving each studio by a separate duct as in Fig. 12-30D.

Heating, ventilating, and air-conditioning contractors in general can handle their part of the design and installation without undue problems. They are usually equipped to prevent or treat problems of noise being introduced into quiet areas by equipment they install. However, noise control usually costs more than a run-of-the-mill installation and the owner would do well to know enough about studio requirements to make good judgments and to lay down specifications adequate to meet the goals, but at minimum cost. One specification which is basic and should be insisted upon from the outset is the maximum air velocity at the outlet grille or diffuser. Experience indicates that this should be no more than 500 cfm in studios used for recording.

NOISE FROM GROUND VIBRATIONS

Let us assume that you are interested in a possible location for your studio but you are concerned about ground vibrations from heavy trucks on a nearby road or possibly from railroad traffic. You understand that such vibrations are readily transmitted to your proposed building structure and that the walls can radiate this sound into your studios. Or perhaps you are interested in space in an unfinished building subjected to suspiciously high ground vibrations. You know that a sound level meter reading in a partially completed building or in a highly reverberant space before partitions are erected will not give an accurate representation of the airborne noise in your finished studios. What can be done to evaluate such a noise hazard in advance?

The British Broadcasting Corporation was faced with such a problem.[11] In the construction of the extension to the BBC Broadcasting House, it was necessary to design structures for isolating their studios from vibrations of an underground train before the structural shell of the building was completed. Acoustical engineers measured the vibration velocity amplitudes of the structural surfaces with a accelerometer, a sound level meter with a special calibrated vibration pickup. By making a few reasonable assumptions they were able to calculate the sound pressure levels to be expected in air near the vibrating surface with a satisfactory accuracy and were able to proceed with their studio design. It is well to be aware that such calculated predictions of sound pressure levels resulting from earth or structure vibrations are possible, but an acoustical consultant will probably be required to carry them out.[12]

NOISE FROM LAMPS

Some disturbing buzzes can emanate from fluorescent lighting equipment, presumably from loose laminations in the starter reactors. If fluorescent lighting is considered for studio illumination, the reactors for all lamps should be located in a steel box outside the quiet area. This requires the running of many extra wires, but the precaution is well worth the extra cost and effort. In fact, this procedure is recommended for critical listening areas as well, such as control rooms, mixdown rooms, etc.

The *triacs* typically used in electronic lamp dimmer circuits are also frequent offenders. They cut off the power-line waveform quite sharply, with the result that noisy spikes with plenty of harmonics are generated. Dimmers should be well shielded, and RF bypass capacitors installed at their input/output lines.

CONDUITS

There is little point in building a 50 dB wall and then perforating it with straight conduit runs between control room and studio such as Fig. 12-31A. The more devious plan of sketch B is much better. After microphone cables are pulled through, the ends of the conduit and the box should be stuffed with glass fiber to reduce sound traveling through the conduit. The same principle applies to electrical power wiring.

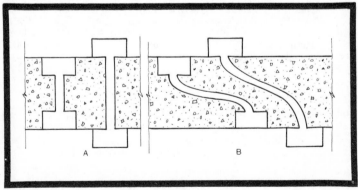

Fig. 12-31. Compromising of high transmission loss wall by improper installation of conduits: (A) worst case; (B) better. Packing of boxes and conduits with glass fiber after wires are in place is advised.

SUPERVISION

During studio construction or renovation there must be close supervision by someone representing the owner who appreciates the vital importance of the points raised in this chapter. This might be the owner himself if he is technically inclined. Most often the studio engineer would be the logical choice. An acoustical consultant could be retained to serve this function or the architect could be so designated. It is almost 100% certain that vital acoustical or functional features will be seriously compromised by well intentioned workmen if this supervision is not given.

REFERENCES

1. Brasch, Jerome. *Vehicular Traffic Noise Near High-Speed Highways.* Sound and Vibration, Vol. 1, No. 12, December 1967 (pp 10—24).

2. Watters, B. *The Transmission Loss of Some Masonry Walls.* Jour. Acous. Soc. Am., Vol. 31, No. 7, July 1959 (pp 898—911).

3. Staff. *A Guide To Selecting Concrete Masonry Walls For Noise Reduction,* published 1970 by National Concrete Masonry Association, Box 9185, Rosslyn Station, Arlington, Virginia 22209.

4. Staff. *Solutions To Noise Control Probems,* publication 1-BL-4589 by Owens-Corning Fiberglas Corporation (1969), Home Building Products Div. Fiberglas Tower, Toledo, Ohio 43601.

5. Consolidated Kinetics Corporation, 249 East Fornof Lane, Columbus, Ohio, 43207.

6. Overly Mfg. Co., Greensburg, Pennsylvania 15602.

7. Rettinger, Michael. *Acoustic Design and Noise Control,* Chemical Publishing Co., New York 1973 (p368).

8. Ferro Cousti-Gasket 1001, distr. by Consolidated Kinetics Corp.

9. Sanders, Guy. *Noise Control in Air-Handling Systems,* Sound and Vibration, Vol. 1, No. 2, February 1967 (pp 8—18).

10. Sanders, Guy. *Silencers: Their Design and Application,* Sound and Vibration, Vol. 2, No. 2, February 1968 (pp 6—13).

11. Gilford, Christopher. *Acoustics For Radio and Television Studios,* IEE Monograph Series 11, Peter Peregrinus, Ltd., 1972 (pp 26, 107, 108).

12. Ward, F. *The Accelerometer as a Diagnostic Tool in Noise Studies in Buildings,* Paper L11, Proc. 4th Intl. Congress on Acoustics, Copenhagen, 1962.

13. Ostergaard, Paul. *Noise Control For Studios,* presented at the 41st Convention of the Audio Engr. Soc., 1971, (Preprint 829G2).

The Budget Multitrack Operation

13

The type of multitrack operation we have been considering is costly. Is there any hope for the impecunious? How about the musician or small musical group, possibly with great talent but of limited resources, who would like to record some demonstration tapes that might lead to a recording contract? How about the technically inclined person who has gone through all the amateur and advanced audiophile stages and has felt frustrated by all the make-do approaches and their built-in limitations of quality in spite of the thousands of dollars he has invested? Or, how about the person, either of musical or technical bent, or both, who would like to get into the recording business eventually, but right now would like to engage in some serious recording as he develops his skill? And there are the many religious musical groups with an eager market for their records, but who simply cannot afford professional studio rates. There are many small recording facilities in basements or garages, and they are growing both in number and sophistication. This is not a very serious threat to the professionals, but rather an asset, for they provide a pool of musical and technical talent contributing to the future growth of the industry.

This book has been prepared primarily for the professional and for those interested in and working toward the field of professional recording. But it is very interesting to note how the amateur audiophile and hi-fi buff have anticipated many of the sound processing techniques now used professionally. Phasing or flanging is an example of this. There is now fancy analog and digital equipment in use in the professional studio to accomplish the effects the resourceful amateur first obtained in his living room from a couple of consumer-type tape recorders used in new and different ways. More power to such explorers of new domains! Who knows where the next good idea is coming from?

SOUND ON SOUND

Before we consider the less expensive equipment which enables those with talent but not much money to achieve truly professional results, it is well for us to review the amateur's time-honored method of doing many of the things normally considered the province of the professional. *Sound on sound*, one of the special effects which have intrigued the audiophile, can be done mono or stereo and on almost any tape recorder. Basically, it is nothing more than an *erase head* defeat. In the usual consumer machine, going to record mode energizes the erase head so that any signals impressed on the tape will be wiped clean before the new signals are recorded. If the erase function is defeated, it is possible to record one signal on top of another already on the tape. The erase can be defeated electrically by throwing a switch which cuts off the erase current, or mechanically by lifting the tape from the erase head. The latter can be easily accomplished by a double thickness of photographic film made into a circle and slipped over the erase head.

Sound on sound affects the original recording. The high frequency bias signal in the record head partially erases the original recording, especially the higher frequencies, but if the original sound is the musical background, reducing its level while the narrator is speaking is required anyhow. The original sound comes to full volume when the *play* button is pushed, removing the bias from the record head. Sound on sound is an effort to mix two or more sounds, something that can be done far better with a simple mixer if both sounds are available simultaneously at the outset.

SOUND WITH SOUND

Sound with sound requires a stereo tape recorder with the proper switching facilities. It is recording one signal on one track and, on a separate run, combining it with another signal on the other track.

Any stereo tape recorder that permits simultaneous playing of one track while recording on the other may be used to add one signal to another as many times as desired. One person can sing one part of a quartet arrangement on one track, play it back through headphones while combining the first track, going back and forth between the two tracks until the one-person quartet is completed. Separate input level controls for microphone and line on each channel actually constitute a simple two-channel mixer. Obviously, a stereo recorder used this way produces a monophonic signal, since the playback of one track prevents its use for recording at that time. But several stereo recorders can be hooked up to achieve the same thing, resulting in a stereo signal after the recording passes are made.

THE PROFESSIONAL APPROACH

Serious multitrack recording requires (a) separation between the sounds of the several performers, (b) a multichannel mixer (usable for both recording and mixdown), (c) a multitrack recorder, and (d) a two-track master recorder. One inexpensive way to get infinite separation is to record the tracks one at a time. Another way to get good separation is to use musical instrument pickups and feed signals from the instrument preamplifiers to the mixing console. There are many situations in which neither of these will do what is needed and one must resort to acoustic separation, which means "the works": treated studios, baffles, isolation booths, or reasonable facsimiles thereof. But there are economic shortcuts and improvisations which can be resorted to even in this area. Someone's living room might serve as a temporary studio, the davenport and some cushions as a baffle for the drums, and so on.

THE BUDGET MULTICHANNEL CONSOLE

In Chapter 3 we looked with covetous eyes at consoles costing from $20,000 to $40,000. There is no substitute for them for the professional job they are built to do. A multichannel console costing but one-tenth of this is possible if one is able and willing to accept fewer channels and some streamlined features. It won't be as flexible, convenient, or awe-inspiring as a Neve 32-channel custom job; but working within its limitations, excellent recordings can be produced and much learned on the science and art of recording.

There are dozens of manufacturers of budget boards, components, and kits which we cannot cover in detail.[1] It will be instructive, however, to look closely at the features of one of them, the Tascam-Teac Model 10 console,[2] to see how much sacrifice is necessary to go to a board in this price class. Figure 13-1 shows the Model 10 in its 8-in/4-out configuration, which can be increased to 12-in/4-out. With a few minor modifications the Model 100 expander board connects to the Model 10, increasing its input capability to 24 input modules. Each input module includes the following:

Fig. 13-1. In this 8-in/4-out console, Tascam Model 10, there is room for 12 input channels. With the Model 100 expander desk (with 12 channels, the total can be increased to 24).

- Input controls
  - Input selector: mike-test-line
  - Switchable mike attenuation: 0−20−40 dB
  - Variable mike attenuator: 0−20 dB
  - Variable line attenuator
- Band limiting filters
  - High-pass: 40 and 100 Hz
  - Low-pass: 5 and 10 kHz
- Equalizer
  - High-frequency: 10 dB dip or boost at 10 kHz
  - Mid-frequency: 10 dB dip or boost at 3 or 5 kHz
  - Low-frequency: 10 dB dip or boost at 90 or 200 Hz
- Echo send
  - Echo send (before or after fader and equalizer)
  - Echo send (off/on switch fader and equalizer)
  - Echo send level control
  - Echo send channel assignment (4 pushbuttons)

- Channel assignment (4 pushbuttons)
- Panpot
- Channel fader (linear)

Each of the four submaster modules (output lines) have the following controls:

- VU meter selection switch: line, echo send level
- Monitor level control
- Monitor selection switch: record, off, play
- Echo receive level control
- Submaster fader (linear)

The master gain module contains only a four-gang linear fader which controls the line output of the board.

The quad panner module has a joystick panpot for each of the four submaster outputs.

The talkback module contains a key-operated microphone and a talkback level control. A 5-watt amplifier is associated with this module for driving the talkback loudspeaker in the studio. When the key is pressed, the control room monitors are muted to prevent howlback.

The last module is the remote control module for controlling all operating functions of the tape recorder: fast forward, rewind, stop, play, and record. Another button on this module controls a warning light or power to a receptacle, to control other devices that should be switched on or off during a recording.

A headphone monitoring panel is on the edge of the board nearest the operator. The VU meters have LED peak indicators which indicate fast peaks to which the meters cannot fully respond.

We note in this inexpensive board a high percentage of the features of the larger boards. True, there is no solo button, no cue mixing facilities (although echo send can be used for a cue mix during the original recording), the equalizer has less detail, no test tone, no jack field, and so on. But you will observe that most of the basic features detailed in Chapter 3 are present in one degree or another and that deficiencies have minimum effect on quality, a greater effect on convenience.

BUDGET MULTITRACK RECORDERS

In the field of multitrack recorders we find that the prevailing situation is quite similar to that in the console field: there are the expensive and the less expensive types available. The budget operation will undoubtedly be limited in the number of tracks, because going to the machines recording 8 or more tracks on 2 in. tape is expensive from the standpoint of both initial cost and equipment operation. Perhaps the best thing to do, with limited funds, is to think initially in terms of a 4-track recorder.

The Tascam Series 70 recorder/reproducer is a suitable companion for the Model 10 board. It is available in 2-track, ¼ in., and 4- and 8-track ½ in. configurations. The unusual approach of recording 8 tracks on ½ in. tape instead of 1 in. tape is an effort to provide the semiprofessional with an

Fig. 13-2. A budget 4-track, ½ in. recorder/reproducer, Tascam Series 70, with 701 electronics. The Series 70 recorder is available in an 8-track, ½ in. version, also with a sync system for overdubbing.

inexpensive 8-track facility which is reasonably economical to operate.

The Tascam Series 70 recorder is available with either of two types of electronics. The Model 501 electronics channel is equipped with mike preamp, standard connectors, and a headphone monitoring output, and it is adapted to interfacing professional equipment. The Model 701 electronics is suited for interfacing the Model 10 board or other high-impedance source. The interconnecting lines should be held to less than 15 ft. The 703 overdub unit contains the sync electronics and the bias oscillator. Figure 13-2 shows a Tascam Series 70 recorder/reproducer (½ in. tape) with 701 electronics.

A 4-track ¼ in. tape recorder/reproducer worth looking into is the Crown CX 844 4-channel unit of Fig. 13-3. This machine is offered by the manufacturer of the well known Crown DC300A monitoring amplifier used widely in recording

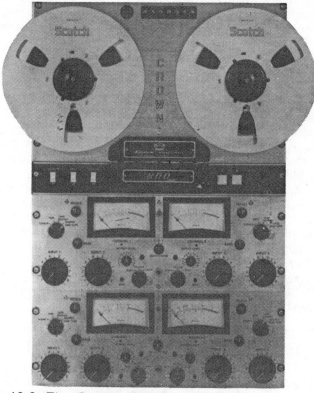

Fig. 13-3. The Crown CX844 4-track, ¼ in. recorder/ reproducer. Overdubbing can be done on this machine with CrownTrac-Sync accessory.

studios.[3] Computer logic prevents tape breakage through foolproof operating controls. A reel-motor switch facilitates editing and a drive motor switch keeps the electronics on with the motor off. A cue lever controls tape-to-head contact and the machine is adaptable to remote control. With the *Trac-Sync* accessory, overdubbing is available. The unit features front-panel bias control, 3-speed equalization, and treble and bass controls with detents for *flat* condition. The frequency response specifications state ±2 dB, 40 Hz−30 kHz, 20 Hz−25 kHz, and 20 Hz−15 kHz for the 15, 7½, and 3¾ ips speeds. Signal-to-noise ratios are 60, 60, and 55 dB for the same speeds. Erase and crosstalk figures of 50 dB are specified. Wow and flutter specifications are 0.5%, 0.09%, and 0.18% at 15, 7½, and 3¾ ips.

When we get closer to consumer products we must distinguish between *4-track* and *stereo 4-track* recorders. In the 4-track recorder there are four independent channels of electronics so that all four tracks may be recorded simultaneously. Not so on the stereo 4-track machines, which have only two channels of electronics and are designed to record a stereo pair on tracks 1 and 3 on one pass, and tracks 2 and 4 on the next pass with the tape turned over.

A number of lower-priced 4-track tape recorders using ¼ in. tape are available in the advanced audiophile class. The Akai GX-400D and the Teac 3340 are examples of these. The Teac 3340S, pictured in Fig. 13-4, has four separate input channels, four quasi VU meters, and it records on 4 discrete tracks. With an external mixer, such a recorder could serve a simple 4-track operation. Each track of the Teac *Simul-Sync* record head can be electronically switched to permit playing back of, say, three of the tracks for cue while recording on the fourth track. This is normal overdubbing. A punch-in feature also allows one to go directly from play to record modes, presumably with no telltale clicks on the track. The wow and flutter specifications (0.04% at 15 ips, 0.06% at 7½ ips), frequency response (±3 dB 30 Hz−30 kHz at 15 ips, 40 Hz−24 kHz at 7½ ips), and signal-to-noise ratio of 55 dB should yield clean basic recordings.

PREMIXING

Limiting the operation to 4 tracks will invariably require some premixing. With a 4-track recorder, an 8- or 12-channel board such as the Tascam Model 10 allows some premixing on the board. For example, 4 drum microphones could feed into 4 input channels which could all be punched into a single submaster bus. When the capacity of the board is exceeded,

Fig. 13-4. A budget 4-track, ¼ in. recorder/reproducer in the advanced audiophile class is this Teac 3340S which features Simul-Sync for overdubbing.

outboard mixers are common to mix several microphones into a single console channel. Premixing is philosophically at odds with the whole concept of multitrack flexibility, but is very commonly practiced as an economic compromise. Of course, premixing must be confined to related components, such as the various mikes on the drum kit.

Premixing can best be done with standard input channels on the console with full equalization facilities available. If there is a shortage of input channels, the next best approach would be to roll in another mixer with equalization control. If there is no spare board available and equalization is not imperative, compact and inexpensive mixers such as the Shure M67 are frequently used.

NOISE REDUCTION FOR BUDGET OPERATIONS

Even if the small recording operation is limited to 4 tracks, tape noise can be a problem, not only the steady hiss so familiar on quiet passages, but particularly the so-called modulation noise that changes with the signal envelope. The signal tends to mask noise in the same frequency region, but strong low-frequency signals can create varying tape hiss in the high-frequency range that can be very noticeable and objectionable.

The DBX company has introduced a line of noise reduction systems that may be suited to the budget operation.[4] They are said to give 30 dB noise reduction, 10 dB more recording headroom, and are compatible with all professional DBX systems, although not with Dolby. No pilot tones are used and there are no complex alignment procedures. In fact, these have the same basic electronics as the DBX professional system described in Chapter 4, except that they have single-ended inputs and outputs rather than the balanced terminations required in professional studios to interface with other equipment.

The Model 157 is illustrated in Fig. 13-5. This unit contains two channels of record and two channels of play and can be used simultaneously for recording and monitoring. Two of these units would allow simultaneous recording and monitoring of four channels. The Model 154 of Fig. 13-6 has 4 channels of noise reduction. They can be switched to record or playback, but not for both simultaneously. The DBX units in

Fig. 13-5. The Model 157 semiprofessional noise reduction system features 2-channel record and 2-channel play for simultaneous recording and monitoring.

Fig. 13-6. The Model 154 semiprofessional noise reduction system has 4 channels which can be switched between record and play.

the 150 series cost somewhat more than half the professional units if they have the simultaneous record and play feature, and considerably less than half if switchable.

Compression and Expansion

In Fig. 13-7 the Model 119 *Decilinear* compressor/expander is shown. This is a wide-range stereo unit designed for both the audiophile and the professional for applications in which extreme compression or expansion is

Fig. 13-7. The Model 119 Decilinear dynamic range compressor/expander. This is a wide-range stereo unit for both the audiophile and the professional where wide-range compression or expansion is required. Compression ratios up to infinity are selectable.

required. Compression ratios up to infinity are selectable and it has a capability of expanding or compressing signals exceeding a preset threshold.

If it is determined that such signal processing is required in the budget recording operation, those devices can do the job at modest cost.

REFERENCES

1. A partial list of these, arranged alphabetically: *Allen and Heath*, distributed by Audiotechniques, Inc. 142 Hamilton Ave., Stamford, Connecticut 06902. *Fairchild Sound & Equipment Corp.*, 75 Austin Blvd., Commack, New York 11725. *Gately Electronics*, 57 Hillcrest Ave., Haverstown, Pennsylvania 19083. *Interface Electronics*, 3810 Westheimer, Houston, Texas 77027. *Opamp Labs, Inc.*, 172 S. Alta Vista Blvd., Los Angeles, California 90036. *Shure Brothers, Inc.*, 222 Hartley Ave., Evanston, Illinois 60204. *Sunn Musical Equipment Company*, Auburn Industrial Park, Tualatin, Oregon 97062.

2. *Teac Corporation of America*, 7733 Telegraph Road, Montebello, California 90640.

3. *Crown International*, 1718 W. Mishawaka Road, Elkhart, Indiana 46514.

4. *DBX, Incorporated*, 296 Newton Street, Waltham, Massachusetts 02154.

Glossary

acoustic response
> Response of a reproducing system including power amplifier, loudspeaker, and room acoustics.

ambience
> Spatially diffused reverberant sound in an enclosure which gives the impression of size and degree of liveness of the enclosure.

analog
> Analog techniques are based on representing signals with proportional electrical voltages or currents. An ordinary amplifier is an analog device, as is a voltmeter. (See **digital**.)

anechoic room
> Literally, a room without echoes. A room whose boundaries absorb the incident sound, thereby affording essentially free-field conditions.

assignment (See **delegation**.)
attack time
> In signal processing devices actuated by signal level (e.g., compressors and expanders), the attack time describes how fast the resulting effect takes place when the signal level is suddenly increased.

attenuator
> An adjustable resistance element in electronic circuits for varying the signal level.

baffle
> Portable sound barrier used in recording studios to provide some degree of acoustical separation between instruments or performers. Also called *screen*. The term is also used for the loudspeaker mounting board which affects its radiation efficiency.

bass

The energy of sound in the lower frequency range of the human ear.

biamplifier

In the biamplified loudspeaker system separate power amplifiers are used for the high- and low-frequency radiators. The dividing network is at low level before the power amplifiers. (See **dividing network; triamplifier**.)

bias

In magnetic tape recording the signal to be recorded is mixed with a high-frequency (60—100 kHz) bias current to reduce distortion in the magnetization process of the tape.

board

A synonym for the control-bearing desktop of an audio control console. (See **desk; console.**)

buffer

An amplifier used for isolating one circuit from another to avoid interaction.

bus

A term designating major electronic circuits to which many other circuits may be connected, usually at high level such as output bus. Synonymous with *line*.

cardioid

Microphone directivity pattern having a shape like a heart.

coloration

When an acoustic or electrical signal is changed by some external effect, it can be said to be *colored*. For example, room acoustics can color a sound emitted in the room. Derived from light terminology; when the spectrum of white light is affected by an outside agent such as a glass filter, the light becomes colored.

comb filter

A filter having a series of passbands and stopbands distributed uniformly down through the audible spectrum like the teeth of a comb.

compressor

An electronic signal processing device for reducing the dynamic range of a signal. When the dynamic range of the sound source (a symphony orchestra may have a dynamic range of about 70 dB) exceeds that of the recording system, it is necessary to employ either manual or electronic compression. (See also **expander.**)

304

console

In audio systems the console is the operational control center which houses the faders, level indicators, control circuits, equalizers, and switches required to function conveniently and efficiently. (See also **desk; board**.)

critical bands

In human hearing, only those components of noise within a narrow band will mask a given tone. This is called a critical band. The critical bandwidth of the ear varies with frequency, but is approximately 1/3 octave in width.

critical distance

In an enclosure filled with sound from a loudspeaker, the critical distance from the loudspeaker is that distance at which the direct (inverse square) sound level equals the general reverberant level. (See also **reverberant field; inverse square**.)

crosstalk

The signal of one channel, track, or circuit interfering with another is called crosstalk (from early telephone usage).

cue (See **foldback**.)

damping

The introduction of a dissipative element into any resonant system. In electrical circuits, resistance provides damping. In an acoustically resonant structure such as a Hemholtz type absorber, glass fiber serves as a damping element.By damping, a resonance peak is reduced and broadened.

dB (See **decibel**.)

dB (A)

A sound level meter reading made with a weighting network simulating human ear response at a loudness level of 40 phons.

dB(B)

A sound level meter reading made with a weighting network simulating the response of the human ear at a loudness level of 70 phons.

dB(C)

A sound level meter reading made with no weighting network in the circuit, i.e., flat. Decibels referenced to a zero-decibel sound pressure level of 0.0002 microbar.

dBm

Decibels referenced to a zero-decibel level of 1 mW in a 600Ω circuit.

dead room (See **anechoic room**.)

decibel (dB)

The human ear responds logarithmically and it is convenient to deal in logarithmic units in audio systems. The decibel is such a logarithmic unit, always referred to some reference level such as 0.0002 microbar for sound pressure, 1 mW in 600Ω in audio circuits. The smallest change in sound level the human ear can detect is 1 to 3 dB, depending on circumstances.

decode

To restore an encoded signal to usable form. (See also **encode**.)

delay

Digital, analog, or mechanical techniques may be employed to delay one audio signal with respect to another. Useful in sound reinforcement systems to compensate for sound transit times, and in recording for special effects such as phasing. (See also **phasing; flanging; comb filter**.)

delegation

Routing of input channels to output lines.

desk

A synonym for *audio control console* in audio systems, more common in British usage. (See also **board; console**.)

digital

Digital techniques are based on the on/off binary language of computers. Analog signals may be translated to digital signals by sampling at regular time intervals and noting the amplitude of each sampling pulse. Digital signals may readily be retranslated to analog form. (See also **analog**.)

distortion

Any change in waveform or harmonic content of an original signal as it passes through a device. This results from a deviation from strict linearity within the device. Distortion generates harmonics not in the original signal.

dividing network

An electrical network dividing the audible spectrum for efficient and faithful sound reproduction. In a three-way loudspeaker the dividing network sends only high-frequency energy to the high-frequency radiator (tweeter), only midrange energy to the midrange radiator, and only low-frequency energy to the low-frequency radiator (woofer).

doppler effect
A source of sound moving toward or away from an observer gives the observer the sensation of a changing pitch referred to as *Doppler shift*. The pitch increases as the source approaches, and decreases as it recedes.

dynamic range
All audio systems are limited by inherent noise at low levels and by distortion at high levels. The usable region between these two extremes is the dynamic range of the system, expressed in decibels.

echo (See **reverberation**.)

effects, sound
The addition of recorded sounds for background or to achieve naturalness or dramatic effects.

effects, special
Signal processing to achieve a desired effect for novelty or to create a desired mood.

encode
Changing a signal in one way to achieve a given end with the expectation of applying a decoding step at a later stage. For example, certain noise reduction systems require encode/decode steps to achieve the desired result. (See also **decode**.)

equalization
Adjustment of frequency response of a channel to achieve a flat or other desired response. Also applied to the adjustment of frequency response of the overall monitoring system, including the acoustics of the room.

erase
Magnetic tape having previously recorded signals on it must be erased before reuse. This is accomplished in recorders by a high-frequency current in an erase head over which the tape passes just before it reaches the recording head. Bulk erasers producing strong alternating magnetic fields erase rolls of tape prior to mounting on the recorder.

expander
An electronic signal processing device for increasing the dynamic range of a signal to restore a compressed signal to its original dynamic range.

fader

An adjustable resistance element in an electrical circuit to control the sound level (volume). (See also **potentiometer; gain control; volume control**.)

figure-8

A microphone directivity pattern resembling a figure 8 having good response front and back and nulls at the side.

filter

An electronic circuit useful for separating one part of the audible spectrum from another. A *high-pass filter* passes only the energy above a certain cutoff frequency, a *low-pass filter* only the energy below a cutoff frequency. A *bandpass filter* rejects energy below and above and passes only energy within a band.

flanging

An ethereal swishing, inside-out sound achieved by mixing a signal with the same signal slightly delayed. The name derives from an early method of achieving the effect with two tape recorders and slowing one by applying pressure to the flange of its supply reel. Commonly produced today by analog or digital delay devices. (See also **fuzz; phasing; comb filter**.)

flux

The lines of force of a magnetic field.

foldback

The signal fed to the headphones of the performer in the studio engaged in overdubbing. (See also **overdub**.)

free field

A sound field in which the effects of the boundaries are negligible over the region of interest.

frequency

The number of signal cycles occurring during each one-second period. Frequency is the physical measurement corresponding to the sensation of pitch. Audible sound falls roughly in the frequency range 20 Hz to 20,000 Hz.

fuzz (See **flanging**.)

gain control

An adjustable resistance element in electronics for controlling signal voltages.

gating circuit

An electronically controlled circuit which, for example, cuts off a circuit when the signal level falls below a predetermined value.

Haas effect

When one is listening in a reverberant enclosure. the sound arriving at the ear within the first 35−40 msec is integrated by the hearing mechanism and interpreted as coming directly from the source. Sound arriving after this period is perceived as an echo or as random reverberation.

harmonics

A harmonic is any multiple of the fundamental frequency.

intermodulation distortion

If a nonlinearity exists in an amplifier, a signal component at one frequency will modulate a signal component at another frequency, creating new components which are called intermodulation (IM) distortion products.

inverse square

Sound pressure falls off inversely as the square of the distance from the source in the absence of all reflecting surfaces (such a condition is approached outdoors). (See also **reverberation field; critical distance**.)

isolation booth

A small enclosure associated with a recording studio in which an instrumentalist, vocalist, or narrator may be isolated acoustically from sounds in the studio. Visual contact is considered desirable.

jack field

Signal paths of console electronics are routed through jacks in the jack field. By plugging into these jacks the normal configuration of the console may be changed for unusual applications or for locating trouble.

lacquer

The master disc recording is cut in a lacquer coating on a very flat aluminum disc. The disc is often referred to as the *lacquer*.

LED

Light-emitting diode, a semiconductor device which emits light when voltage is applied.

level

The level of a quantity is the logarithm of the ratio of that quantity to a reference quantity of the same kind.

level indicators
Any visual device for indicating signal levels in the various channels of the console or tracks on the recorder. VU meters, dancing bars on a CRT screen, or stacks of light-emitting diodes may be used to indicate level. They may indicate instantaneous signal peaks or RMS (root-mean-square) effective values.

limiter
An electronic signal processing device by which the upward excursion of signal level may be limited to prevent overdriving of following devices.

linear
If the output of a device bears a constant ratio to its input, the device is said to be linear.

live room
A room characterized by reverberation resulting from unusually small amounts of sound absorption.

logic
The sequential and operational steps expressed in electronic circuits of the type required to accomplish a desired function.

mass law
The ability of a barrier to attenuate sound depends upon the mass of the barrier. Transmission loss computed from this simplified mass law may vary considerably from actual measurements.

master
A master tape is the original tape on which is recorded the signals mixed down from the multitrack tape, contrasted to a copy or dubbing (rerecording) of the master tape. The master recorder is the machine on which the master tape is recorded. The master fader controls the level of all console output signals without disturbance of channel faders.

matrix
An electrical network; usually a large number of components of a given type, as resistors. With reference to quadraphony, the *matrix* is the decoding network that reconstitutes four channels from two.

microphone, contact
A special microphone affixed to a musical instrument in such a way that it responds primarily to the vibrations of

the instrument rather than to airborne sounds. Also called *pickup*.

mid-side (Blumlein pair)
A combination of two microphones in one housing for stereo use, one having a cardioid (mid) and the other a figure-8 directivity pattern (side).

mixdown
The process of combining the several signals on a multitrack magnetic tape with appropriate levels, signal processing, and time relationships to give a complete performance in fewer channels than the master contains.

mixer
An electrical or electronic device capable of mixing (summing) two or more signals into a composite signal.

modulation noise
A type of noise generated in magnetic tape, the magnitude of which is proportional to signal level. In most cases the higher signal masks the increased noise but the fluctuating hissing noise produced by heavy bass signals can often be distracting.

monitor
The means for hearing and checking signals for control and evaluation. The name often applies to the loudspeakers in the control room.

monophonic (mono)
Recording and reproduction of sound by a single channel.

multitrack recording
Recording a multiplicity of audio tracks on magnetic tape. The number of tracks ranges from 2 to 40.

octave
The interval between two frequencies having a basic frequency ratio of 2:1.

omnidirectional
Having a uniform response to sound arriving from any direction, as in *omnidirectional microphone*.

operational amplifier (opamp)
An amplifier (usually an IC) requiring a modest number of exterior components to produce a wide range of amplifier characteristics.

overdub
In separation recording it is not necessary that all tracks be recorded concurrently. Individual tracks may be recorded or redone at a later time or even built up one

track at a time. Synchronism is achieved through cue (foldback) signals of previously recorded tracks fed to the performer through headphones. To eliminate a time delay, such foldback signals are picked up from the record heads which are in sync with the track being recorded.

overtones
Overtones bear an octave relationship to the fundamental frequency. The overtones of 100 Hz occur at 200, 400, 800...etc. Hz. Not to be confused with harmonics. (See also **harmonics**.)

panoramic potentiometer
An adjustable resistance network in an electric circuit by which the signal on a given channel may be positioned in the stereo or quad playback sound field.

panpot (See **panoramic potentiometer**.)

phantom image
By phase and level adjustments the apparent source location in stereo or quad may be shifted between two loudspeakers with little relationship to the true source location. This apparent location of the source is called the phantom image.

phase
The time relationship between two signals.

phasing
The term applied to the proper connection of loudspeaker or microphone leads so that all elements are properly in phase. (See also **flanging, comb filter**.)

pink noise
A noise signal whose spectrum level increases 3 dB per octave. It is convenient to use pink noise with an analyzer having a bandwidth of a given percentage of the frequency to which it is tuned (e.g., 1/3 octave). (See also **white noise; random noise**.)

potentiometer
An adjustable resistance element in an electrical circuit by which magnitudes of voltages and currents may be controlled. (See also **fader; volume control; gain control; attenuator**.)

power amplifier
An amplifier designed to supply relatively large amounts of power to drive loudspeakers, etc.

preamplifier
> An amplifier optimized for low noise to amplify low-level signals such as those from microphones.

precedence effect (See **Haas effect**.)

premixing
> Employing an outboard mixer to mix signals from several related microphones before sending the combined signal to the main console. A device for overcoming the problem of limited input channels in the main console.

program amplifier
> An amplifier used at intermediate signal levels between the preamplifiers and power amplifiers.

quadpot (See **panoramic potentiometer**.)

quadraphony
> The recording and reproduction of sound through four channels to give more spatial and ambient information than is possible with only two channels. The adjective form is *quadraphonic*.

random noise
> An oscillation whose instantaneous magnitude is specified only by a probability function—useful in acoustic measurements.

release time
> In signal processing devices actuated by signal level (e.g., compressors and expanders) the release time is a measure of the time required for the device to be restored to normal operation upon sudden removal of the actuating signal.

remix (See **mixdown**.)

reverberation field
> The reverberant sound or reverberation field in a room is that built up by sound which has undergone multiple reflections. The level of the reverberant field is essentially constant throughout the room. (See also **critical distance; inverse square**.)

reverberation
> When any enclosure, recording studios included, is filled with sound and that sound is suddenly cut off, a finite length of time is required for the sound to reach inaudibility. This tailing off of sound resulting from multiple reflections from the boundaries of the enclosure is called reverberation.

reverberation, artificial

Artificial reverberation is commonly added to audio signals to simulate the ambient conditions of a large hall. The delayed effect required in generating artificial reverberation may be produced in a reverberation chamber having hard, reflective surfaces, or by sending sound waves down a metal spring, or within a metal plate. Digital or other delay lines can be used in producing artificial reverberation as well as magnetic tape or discs.

rolloff

A gradual decrease in response of a circuit as contrasted to terms *cutoff* or *chopoff*, designating faster decrease in response.

saturation

In all magnetic tape there is a point above which the output is not proportional to the input. This is the region of magnetic saturation (to be avoided in recording).

screens (See **baffles**.)

separation recording

The technique of fragmenting a musical group and recording each component with a separate microphone (or group of microphones), a separate channel through the console, and to a separate track on a multitrack recorder. Mixing these tracks in a later operation gives the final performance.

servo

A servo system is a device controlled by sensing the output. For example, let us say we want to control the speed of a motor. The motor may be driven by a servoamplifier controlled by a speed sensor attached to the motor shaft so that deviations from the desired speed produce error signals which compensate for the deviations. It is a feedback system.

shelving

Instead of a peak or dip in channel response near the high or low end of the audible band, an equalizer may also go up or down to a certain level and remain flat at that level. This flat region is called a shelf, and the circuits are referred to as *shelving*.

shift register

An integrated circuit of great complexity used for achieving time delay for audio use.

signal-to-noise
>   The number of decibels separating the signal level from the noise level of a channel, magnetic track, or electronic device.

slating
>   Recording voice identification signals on the magnetic tape at the beginning of a take.

sound absorption
>   Sound energy dissipated as heat in the interstices of fibrous materials or the fibers of flexural panels. The sound absorption coefficient represents that fraction of incident sound absorbed by the material. A coefficient of 0.22 means 22% of incident energy is absorbed.

sound level
>   A sound pressure expressed in decibels as a ratio to some reference level. In acoustics, the accepted reference level is 0.0002 microbar, which is close to the human threshold of hearing.

sound with sound
>   An amateur attempt at overdubbing. Requires a stereo recorder that permits playing of one track while recording on the other. Tracks can be combined successfully as headphone monitoring is used.

sound transmission class
>   A standard method of rating airborne sound transmission performance of a wall or floor—ceiling structure at different frequencies by means of a single number.

standing wave
>   The air in a room constitutes a complex acoustical system having many resonance points. For example, two parallel walls 20 ft apart are acoustically resonant at 28 Hz and multiples of 28 Hz. When sound from a loudspeaker excites this mode, nodes and antinodes of sound pressure, called a *standing wave*, result.

stereophonic
>   Recording and reproduction of sound through two channels to give spatial information, retaining essentially the positional information of the original performance.

synthesizer
>   A device for producing a wide variety of sounds useful for special effects, generating new-sound music, and imitating sounds of conventional instruments and recognizable noises (as clapping, whistling, etc.).

talkback

The facility by which the console operator can talk to those in the studio. Those in the studio can communicate with the control room over the normal microphone channels.

tape hiss

The electrical noise produced by the granularity of the oxide coating of the magnetic tape.

test tones

A test oscillator built into the console generates sine-wave tones of adjustable frequency for conforming console levels with recorder levels, checking frequency response, etc.

three-way loudspeaker

A three-way loudspeaker utilizes three separate radiators, one for high frequency, one for the midrange, and one for the low-frequency components of the signal. (See also **dividing network**.)

transducer

Any element that changes one form of energy to another. Electroacoustic transducers change acoustic energy to electrical or vice versa. Microphones and loudspeakers are such transducers.

transients

A short high-amplitude variation in signal level. Short-lived impulsive bursts of sound pressure or voltage or current in electrical circuits.

transmission loss

As applied to sound transmission through partitions, baffles, or other sound barriers, transmission loss is measured by the number of decibels that sounds are attenuated in passing through the barrier.

treble

The energy of sound in the upper range of frequencies perceived by the human ear.

triamplifier

In the triamplification system, separate power amplifiers are used to drive the high-frequency, midrange, and low-frequency radiators in the loudspeaker; the dividing network is at low level before the power amplifiers. (See also **dividing network; biamplifier**.)

two-way loudspeaker

A two-way loudspeaker utilizes separate high- and low-frequency radiators. (See also **dividing network**.)

voltage-controlled amplifier

An amplifier whose gain can be controlled by varying a direct-current voltage. Used extensively in automated systems.

volume control

An adjustable resistance element in an electrical circuit to control sound level (volume).

VU

Decibels as indicated by the standard VU meter having standardized ballistic characteristics and a standard calibration based on 0 VU = 1 mW of power in a 600Ω circuit. Vocal or instrumental signals are highly transient in nature and the VU meter follows such changes in a standardized way.

wavelength

The distance a sound wave travels in the time it takes to complete one cycle.

white noise

A noise whose noise power per unit frequency is independent of frequency over a specified range. The spectrum of white noise measured with an analyzer having a constant bandwidth is uniform with frequency. (See also **random noise; pink noise**.)

X-Y microphones

A coincident pair of microphones with cardioid or figure-8 directivity patterns used for stereo pickup, arranged so that one picks up the left predominantly and the other the right.

Index